American Institute of Chemical Engineers

Conference on Engineering Sciences and Fundamentals 2007

Held at the 2007 AIChE Spring National Meeting

April 22-27, 2007
Houston, Texas, USA

Printed from e-media with permission by:

Curran Associates, Inc.
57 Morehouse Lane
Red Hook, NY 12571
www.proceedings.com

ISBN: 978-1-60423-362-9

ISBN: 978-1-60423-362-9

American Institute of Chemical Engineers
Proceedings
Three Park Avenue
New York, NY 10016-5991
Phone: 212-591-8100

www.aiche.org

American Institute of Chemical Engineers

Conference on Engineering Sciences and Fundamentals 2007

Held at the 2007 AIChE Spring National Meeting

AIChE

April 22-27, 2007
Houston, Texas, USA

American Institute of Chemical Engineers

Conference on Engineering Sciences and Fundamentals
2007

TABLE OF CONTENTS

Session 10 – Prediction and Correlation of Transport Properties

Session 37 – Thermodynamics and Transport Fundamentals

Session 60 – Thermodynamics and Phase Equilibria I

Session 80 – Thermodynamics and Phase Equilibria II

Marangoni Effect and "Chasing Drops"

Preeti S. Yadav, Prashant Bahadur, and Rafael Tadmor. Chemical Engineering, Lamar University, Beaumont, TX 77710

The spontaneous motion of a liquid droplet due to the placement of another drop of a different liquid has been studied. The motion is recorded when both the drops are placed on one smooth surface as well as when the two drops are kept on different surfaces still retaining the same distance between them. It is traditionally believed that Marangoni Effect is caused when vapors of one liquid deposit on the closer region of the other drop thus creating surface tension gradient within the drop. Through our experiments, we observed that despite keeping the same distance between the drops, thereby allowing for vapor deposition, upon surface separation the pursued drop did not run away spontaneously the way it did for continuous smooth surface. This indicates that the surface is an important factor in the spontaneous drop flow.

Emulsification Via Dynamic Interface Transport

Zhengdong Cheng[1], Dawei Luo[2], Sussanna Wong[3], Manuel Marquez[4], and Guangnan Meng[3]. (1) Texas A&M University, Chemical Engineering Department, College Station, TX 77843, (2) Chemical Engineering, Texas A&M University, College Station, TX 77840, (3) Artie McFerrin Department of Chemical Engineering, Texas A&M University, College Station, TX 77845, (4) Arizona State University; National Institute of Standards and Technology; Los Alamos National Laboratory

A novel concept of emulsification is presented through a controlled first order transition in combination with the dynamical interfacial transport and microfluidics. The evaporation of emulsion drops inside an ambient fluid with surfactants triggers nucleation and growth of solute particles in a shell near the interface between two immiscible fluids. Solute particles transport through the interface after growth. The uniformity of the produced particles crucially depends on the uniformity of the emulsion drops.

Interaction of Pressure Sensitive Adhesive Thin Films with Moisture

Haiyan Li[1], Larry E. Gwin[2], Carl J. Houtman[3] and Steven J. Severtson[1]
[1]Department of Bioproducts and Biosystems Engineering, University of Minnesota, 2004 Folwell Avenue, St Paul, MN 55108
[2]Polymer Group, Franklin International, 2020 Bruck Street, Columbus, OH 43207
[3]USDA, Forest Products Laboratory, One Gifford Pinchot Drive, Madison WI, 53726

Pressure sensitive adhesive (PSA) films cast from acrylic water-based formulations interact strongly with moisture. This results in a substantial reduction in their performance properties and aesthetics of label laminates. The problem is most severe for water-based PSAs because of their composition and the greater porosity of formed films. In addition to the adhesive polymers, which possess a small content of polar functional groups, acrylic water-based PSAs contain various amphiphilic components, which are used to emulsify monomers during polymerization reactions and formulate adhesive emulsions for coating. A dried acrylic film will contain as much as 8% surfactant. The greater porosity of water-based films is a result of the way they are generated. These PSAs are made by first coating formulations onto a substrate, typically a release liner, followed by drying during which latex particles coalesce to form an adhesive film. Although it appears there is little published on this topic, the extent of the coalescence process likely varies with the composition, mechanical properties and even structural characteristics of the latex particles. Thus, identifying how changes in the synthesis and processing influence the extent to which moisture moves into and through (i.e., permeability) a water-based PSA is complicated by the fact that these changes can influence both the chemical composition and morphology of films.

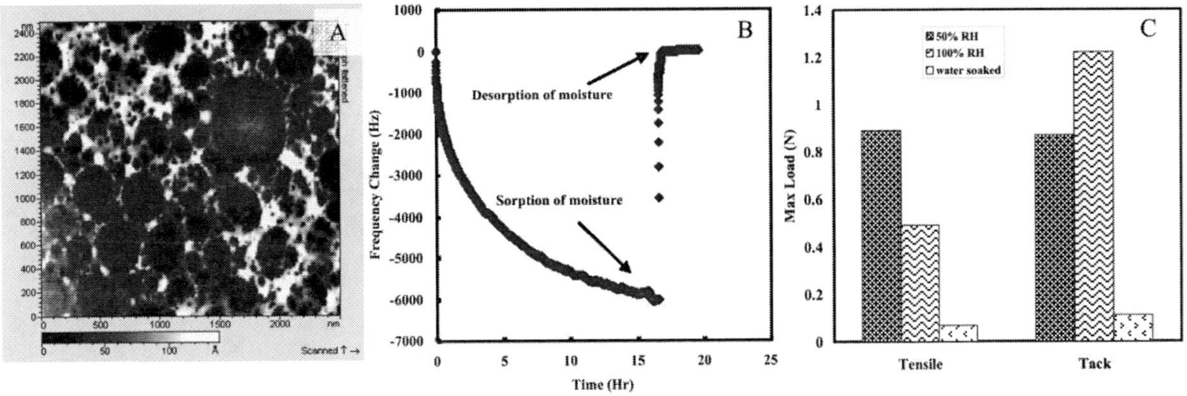

A.) An atomic force microscopy (AFM) image of a commercial water-based acrylic PSA captured in tapping (AC) mode. B.) Frequency shift for a QCM crystal that was coated with this PSA and exposed to 100% (sorption) and then 0% (desorption) relative humidity as a function of time. C.) Various performance properties of the commercial PSA measured at 50 and 100% relative humidity and after the film was soaked in water. Strength is the maximum force required to produce the failure or yielding of the film in tension and tack is the maximum force recorded for a probe tack test run using an 11.1 mm diameter probe with an 8.8 mm radium end.

In this presentation, current research from a study aimed at identifying the dominant variables and mechanisms controlling the sorption of moisture by PSA films will be reviewed. The adhesives examined were model formulations designed to provide a broad range of properties and behaviors based on initial studies with commercial products. Adhesive polymers were produced via emulsion polymerization using the same surfactant system and synthesis scheme. Soft monomers *n*-butyl and 2-ethylhexyl acrylate were combined with lesser amounts of the hard monomers vinyl acetate, styrene and methyl methacrylate and function monomers acrylic acid and methacrylic acid. The focus will be on measurements made using quartz crystal microbalance (QCM) technology under a variety of humidity conditions including cycling. Changes in film morphology and porosity were monitored using scanning probe microscopy techniques and through measures of permeability made using house-built experimental equipment. Data are also presented on the changes in performance properties produced by the moisture sorption. Reviewed data indicate that specific monomer combinations result in films with performance properties that readily degrade when exposed to moisture. The rate and extent of sorption varies with composition, but film morphology also plays a key role. This morphology is determined in large part by the latex characteristics. It was also found to be determined in large part by previous moisture exposure, which is seen indirectly through changes in density, porosity and strength. The reversibility of modifications produced through moisture exposure appears to be highly dependent on the initial film structure. The results advance our understanding of the mechanisms via which moisture interacts and influences the properties of water-based PSA films. This is important to the design of more robust products which can withstand a variety of environments and to the development of new products specifically tailored for high humidity environments.

Modeling the Surface Tension of Pure Fluids Using the Gradient Theory and a Simple Ceos

Ricardo Macías-Salinas, ESIQIE-Departamento de Ingeniería Química, Instituto Politécnico Nacional, UPALM, Av. IPN s/n, Col. Lindavista, México, D.F., Mexico, Manuela Díaz-Cruz, ESIQIE-Departamento de Ingeniería Metalurgica, Instituto Politécnico Nacional, UPALM, Av. IPN s/n, Col. Lindavista, México, D.F., Mexico, and Fernando García-Sánchez, Laboratorio de Termodinámica, Programa de Ingeniería Molecular, Instituto Mexicano del Petróleo, Eje Central Lázaro Cárdenas 152, México, D.F., Mexico.

The well-known gradient theory of fluid interfaces (GTFI) was combined with a simple cubic equation of state (CEoS) to accurately compute the surface tension of non-polar and polar fluids within a wide range of temperatures: from the triple point up to the critical region. For the majority of the pure fluids under study, a remarkably good representation of experimental surface tensions was obtained near their critical points even though the homogeneous-fluid portion of the GTFI was modeled using a simple CEoS that usually fails to properly describe the fluid phase behavior near the critical point. In an attempt to compensate for this weakness exhibited by the CEoS within the GTFI framework, a new scaling functionality was introduced in this work for the influence parameter of the inhomogeneous fluid. The proposed expression for the influence parameter, which carries the information on the molecular structure of the interface, proved to be highly satisfactory along with the Soave or Peng-Robinson CEoS during the representation of experimental surface tensions of various pure fluids including some normal paraffins, isoparaffins, alcohols, aromatics, ethers, CO2, water, acetic acid, among others.

Relating Contact Angles, Drop Size and Line Energy

Preeti S. Yadav, Prashant Bahadur, Kumud Chaurasia, and Rafael Tadmor. Chemical Engineering, Lamar University, Beaumont, TX 77710

The relation between drop radius, r, the force to slide it, and the advancing and receding contact angles, θ_A and θ_R, has been studied. To keep the line energy (energy per $2\pi r$) independent of r, the modified Young equation predicts that θ_A and θ_R change considerably with r. As shown by many investigators, θ_A and θ_R change negligibly, if at all, with r. We show why the modified Young equation is correct and still θ_A and θ_R should hardly change with r. Our results suggest that the Laplace pressure is a significant parameter in inducing the line energy.

Nanoscale Fluid Flow - Molecular Dynamics Simulations with Solid-Liquid Interfaces

Bohung Kim, Mechanical Engineering, Texas A&M University, College Station, TX 77843, Tahir Cagin, Chemical Engineering, Texas A&M University, College Station, TX 77843, and Ali Beskok, Aerospace Engineering, Old Dominion University, Norfolk, VA 23529.

For the cases of flows with very high shear rate, or in the case of dimensions on the order of nm, the continuum or Newtonian hypothesis breaks down. Molecular dynamics (MD) method emerges as a viable approach to address fluid flow problems. MD potentially can address the issues such as solid-fluid interfaces/interactions arising in the nanoscale-regime. Furthermore, large scale simulations based on atomistic models may also address the transition from atomistic to continuum scale approaches. In this study, we run two different sets of MD simulations to elucidate these issues. First, we analyze the size and time averaging effects by performing large size periodic boundary condition MD simulations to evaluate the convergence of average velocity, temperature, and density corresponding to the length of averaging time and size of spatial bins. Based on the convergence, we observe that the error in fluid properties is within an acceptable range in spite of the discrete behavior of the molecules. Second, most previous MD simulations of nano-scale, shear driven fluid flows use specular type walls, consisting of wall molecules with infinite mass. However, the thermal equilibrium at the wall-fluid interface is often neglected in the system. To consider the thermal equilibrium at the wall-fluid interface, we have used the crystal wall model with atomic bonding stiffness and assume that the wall interacts with the fluid molecules and exchanges momentum. Therefore, the net energy of fluid is conserved by exchanging energy with the wall-fluid interface. Hence, our model is a more accurate description of the real physics of the problem, without the need for an imaginary external heat bath connected to the fluid. Our model, with wall-fluid interactions matches the results of existing models. However, since our model implements the energy exchange through the wall-fluid interface, we observe the temperature jump at the wall-liquid interface, which is not possible with previous models.

Structure and Dynamics of Water within Swcnt and Cyclic Peptide Nanotubes

Tahir Cagin, Lijun Liu, and Jennifer A. Carvajal. Chemical Engineering, Texas A&M University, College Station, TX 77843

Nanoscale confinement effects reveal itself as a common theme in areas such as separation of multicomponent molecular mixtures, transport through cell-membranes, applications of nanofluidic devices, etc. Conventionally molecular dynamics methods have been used to address fluid behavior in the nanometers length scale. Here we have chosen two physically well defined systems: single wall carbon nanotubes with various diameter and chirality and self assembled cyclic peptide nanotubes for representing nanoscale confinement for studying water structure and dynamics. Both of these systems present a uniform environment apart from the distinct charactrer of their specific interactions with water. We will report on the effects of environment, diameter and chirality in case of SWCNT.

Diffusion of Protons in Nanopores

Pedro Derosa, Bharat Kolan, and Scott Gold
Institute for Micromanufacturing, Louisiana Tech University

Introduction

Fuel cell related research has experienced a boom in recent years due to the need for more efficient and environmentally friendly energy sources. Miniature and microscale fuel cells are being developed, and in some cases beginning to be commercialized, for portable electronics such as cellular phones and laptop computers as well as for on-chip applications such as power for microelectromechanical systems (MEMS) and chip scale chemical, biological, and environmental sensors. One reason for the attractiveness of fuel cells for portable applications is their high theoretical energy density when compared with traditional battery technologies.

There is a wide array of fuel cell technologies available, including alkaline, phosphoric acid, molten carbonate, and solid oxide fuel cells, each with various advantages and disadvantages. For automotive, portable, miniature and micro-scale applications, the fuel cell technologies that have received the most attention and shown the most promise are the proton exchange membrane (PEM) family, including hydrogen PEM fuel cells, direct methanol fuel cells (DMFCs), and direct formic acid fuel cells (DFAFCs). In these fuel cells, the fuel is oxidized at the anode to yielding protons, electrons, and other reaction products. Electrons pass through a circuit to power a load while protons pass through the PEM to participate in an oxygen reduction reaction at the cathode.

Porous materials have been the subject of several recent studies, being used primarily as structural components in a PEM. One of the most notable examples of this approach is the Gore® membrane in which the pores of a porous Teflon® membrane are filled with Nafion®.[1] Several other studies have utilized porous silicon, filling the pores with Nafion® and other ionomeric materials with some success. In a similar vein, porous alumina has been filled with solid acid materials for use as a PEM.[2,3] The ionic conductivity of porous ceramic materials has been a subject of interest for many years, though mainly for sensor applications.[4,5] The bulk proton conductivities of metal oxides, clays, and other ceramics are several orders of magnitude lower than that of Nafion®[6] which has led many to dismiss such materials from consideration as potential PEM materials.[7] While proton conduction in the bulk solid of ceramic materials is indeed very slow at temperatures of interest for PEM fuel cells, proton conduction along the acidic surfaces of these materials can be orders of magnitude faster.[4,8]

A number of theoretical works have addressed the issue of proton diffusion. Serowy et al.[9] reported a simulation study of proton diffusion along lipid layers. Proton diffusion is observed to be delayed in protein surface because of the proton binding sites on the surface. In this work Monte Carlo simulation in a cubic spatial grid was conducted by placing the proton randomly in the bilayer and applying infinite random walk in the direction parallel to the membrane and limited in direction perpendicular to the membrane. Proton diffused through the lipid surface and reached aqueous phase with a diffusion coefficient of $5.8\times10^{-5}cm^2/sec$ which is slower than diffusion in bulk water ($9.3\times10^{-5}cm^2/sec$[10]). Their results indicate the existence of a kinetic barrier for proton transfer from bilayer surface to bulk water.

Morrone et al.[11] reported a simulation study of proton transport dynamics in methanol-water solutions. Ab Initio Molecular Dynamics (AIMD) simulations were used to study the structure of methanol-water mixtures at two different mole fractions and proton diffusion at mole fraction of 0.5. Study of proton diffusion in partially aqueous environment like methanol-water mixtures gives idea of how the proton diffuses in biological systems and membranes. The authors claimed that transport of proton in bulk water is driven by structural diffusion. Proton transfer may occur in water-water, methanol-water, water-methanol and methanol-methanol combinations and the authors calculated the occurrence of all these as 36%, 32%, 27% and 5% of the cases respectively, which implies that methanol-methanol is poor in proton transport. The diffusion coefficient of 4.2×10^{-5} cm^2/sec was found for proton diffusion in the mixture and $6.69 \times 10^{-5} cm^2/sec$ for deuterium in heavy water.

Methodology

Molecular dynamics is a computer simulation technique based on the solution of the equations of motion for interacting atoms. During the simulation, Newton's Equations of Motion are solved for all the atoms in the structure and their positions and velocities at each step are obtained. NAMD (No Another Molecular Dynamics) simulation software is used to study the dynamics of the proton diffusion in bulk water and confined space (nanopores of different dimensions). Amber force field is used for these simulations. In all the simulations NPT ensemble is used, pressure is set to one atmosphere and temperature to 300 K.

Results

Diffusion of protons in nanopores was studied with and without electric field. For diffusion without electric field, the simulations were run for 100 ps. Diffusion coefficients were calculated from the Mean square displacement as described below.

$$D = \frac{1}{6}\frac{d\left\langle (r(t) - r(0))^2 \right\rangle}{dt} \qquad [1]$$

For diffusion in an electric field a charged slab was added to the simulation box. Changes in electric field were simulated by changing the charge on the slab. Diffusion under an electric field was simulated inside a nanopore with and without water and compared to diffusion in bulk water.

Diffusion in Nanopores

In order to minimize the number of atoms in the simulation box and thus reduce the simulation time, a straight nanopore was created as a tube-like structure, however no water or protons are allowed in the outside of the pore and the outer wall is kept fixed. Admittedly the structure is too simple but it is a necessary step to study the process.

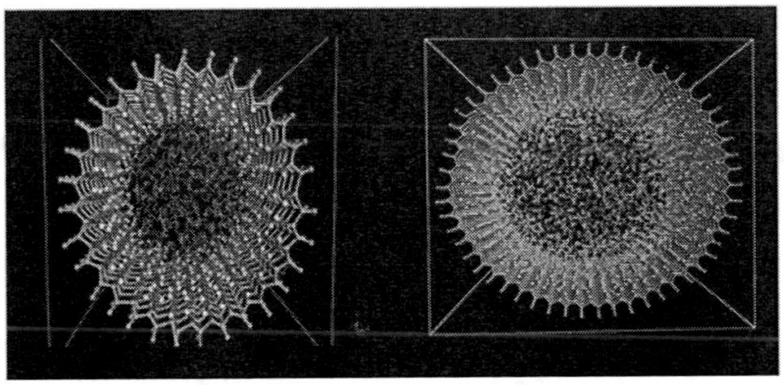

Figure 1: snapshot of the diffusion of a single proton in a) 2.5nm-diameter pore and b) 5nm-diameter box.

The study of diffusion through nanopores of different shapes is currently in process and will be reported elsewhere. Figure 1 shows two pores of different size, 2.5nm and 5nm in diameter with 1 proton inside. These structures were used to study diffusion of protons in constraint space. The simulation time was 100 ps and the diffusion coefficients were calculated to be $D=1.33\times10^{-7}cm^2/sec$ and $D=3.41\times10^{-7}cm^2/sec$. Available data on proton diffusion in bulk water reports a diffusion coefficient of $D=9.3\times10^{-5}cm^2/sec$.[10] i.e. more than 2 orders of magnitude bigger. Our conclusion is that constraint actually delays diffusion. This hypothesis is now being tested by a systematic study of proton diffusion in increasingly large nanopores.

Diffusion in the Presence of an electric field

For fuel cell applications, the effect of an electric field on proton diffusion is of relevance since the electric field is the driving force in such an environment. We studied diffusion under an electric field in a nanopore with and without water (to study the role of water in the diffusion) and we compared with diffusion in bulk water in the presence of an electric field.

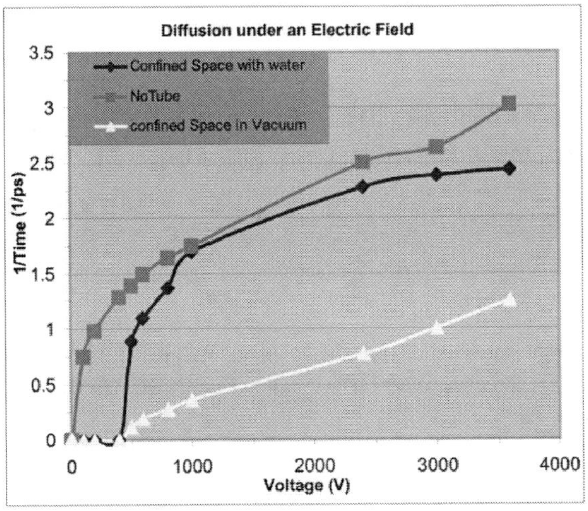

Figure 2: snapshot of the diffusion of a single proton in a) 2.5nm-diameter pore and b) 5nm-diameter box.

Figure 2 and table 1 show comparative results on the diffusion of protons under an electric field. The results in Figure 2 are reported as 1/t vs. voltage. 1/t is reported since for small electric field, the proton does not progress through the pore. These cases are represented with 1/time=0. The voltage is between the slab and the top of a 32nm long pore.

Results show that diffusion through water is faster than in vacuum, this is expected since it is assumed that protons diffuse forming hydronium H_3O^+ rather than alone. Diffusion seems to be slowed down by confinement. Simulations in larger tubes are needed to confirm this fact. A rather large voltage is necessary for the proton to diffuse, this is evidence of the effect of the nanopore surface.

Table 1: Diffusion time vs. applied voltage. Comparison between diffusion through a nanotube, with and without water, and in bulk water. ND implies that the proton just move around the initial position but it does not progress towards the charge slab within the simulation time (10ps)

	Diffusion Time (ps)		
Voltage (V)	**21x32 nm Pore with Water**	**Pore in Vacuum**	**Bulk Water**
1	ND	ND	ND
5	ND	ND	ND
10	ND	ND	ND
100	ND	ND	1.34
200	ND	ND	1.02
400	ND	ND	0.78
500	1.13	8.40	0.72
600	0.91	5.40	0.67
800	0.73	3.70	0.61
1000	0.59	2.80	0.57
2400	0.44	1.30	0.40
3000	0.42	1.00	0.38
3600	0.41	0.80	0.33

Conclusions and Future work

The use of nanoporous membranes proton exchange membranes for fuel cells is increasing but still there is no clear understanding of the actual mechanism of proton diffusion.[12] The results reported in this proceeding are the first steps towards a systematic study of diffusion in nanopores. They should be taken as qualitative at this point but a good agreement with known or expected behavior of diffusing protons is observed.

Our results show that proton diffusion is indeed mediated by the solvent but it is slowed down when in confined space. The slow down of the diffusion in constrained space is observed in both in electric field driven and free diffusion.

It is also observed that the charged inner wall of the nanopore strongly interact with the proton delaying the diffusion, at least for the sizes studies here, larger voltages are then necessary to force the proton through the pore.

Several follow up simulations are required to have a better picture of this process. Different solvents will be tested. Given the role of the solvent in the diffusion of proton, choosing and adequate solvent may prove an important step in the design of PEM for fuel cell applications. Other pore sizes need to be tested to determine the extent if the proton-wall interaction. Different surface chemistry will also be tested for the same purpose; in addition it will help our attempts to produce a model to assist in the design of nanoporours membranes for proton exchange membranes.

References

(1) Kerres, J. A., Development of ionomer membranes for fuel cells. *Journal of Membrane Science* **2001,** 185, (1), 3-27.

(2) Bocchetta, P.; Chiavarotti, G.; Masi, R.; Sunseri, C.; Di Quarto, F., Nanoporous alumina membranes filled with solid acid for thin film fuel cells at intermediate temperatures. *Electrochemistry Communications* **2004,** 6, (9), 923-928.

(3) Park, Y.-I.; Nagai, M.; Kim, J.-D.; Kobayashi, K., Inorganic proton-conducting gel glass/porous alumina nanocomposite. *Journal of Power Sources* **2004,** 137, (2), 175-182.

(4) Knauth, P.; Tuller, H. L., Solid-state ionics: Roots, status, and future prospects. *Journal of the American Ceramic Society* **2002,** 85, (7), 1654-1680.

(5) Montesperelli, G.; Pumo, A.; Traversa, E.; Gusmano, G.; Bearzotti, A.; Montenero, A.; Gnappi, G., Sol-gel processed TiO2-based thin films as innovative humidity sensors. *Sensors and Actuators, B: Chemical* **1995,** B25, (1-3), 705-9.

(6) Poinsignon, C., Protonic conductivity and water dynamics in swelling clays. *Solid State Ionics* **1997,** 97, (1-4), 399-407.

(7) Ruiz-Hitsky, E.; Galvan, J. C.; Merino, J.; Casal, B.; Aranda, P.; Jimenez-Morales, A., Proton conductivity in Al-montmorillonite pillared clays. *Solid State Ionics* **1996,** 85, (1-4), 313-317.

(8) Vichi, F. M.; Colomer, M. T.; Anderson, M. A., Nanopore ceramic membranes as novel electrolytes for proton exchange membranes. *Electrochemical and Solid-State Letters* **1999,** 2, (7), 313-316.

(9) Serowy, S.; Saparov, S. M.; Antonenko, Y. N.; Kozlovsky, W.; Hagen, V.; Pohl, P., Structural Proton Diffusion along Lipid Bilayers. *Biophysical Journal* **2003,** 84, 1031-1037.

(10)Pines, P.; Antonenko, Y. N.; Yaguzhinsky, L. S., Geminate Recombination in Excited-State Proton-Transfer Reactions - Numerical-Solution of the Debye-Smoluchowski Equation with Backreaction and Comparison with Experimental Results. *Journal of Chemical Physics* **1990,** 88, 5620-5630.

(11)Morrone, J. A.; Haslinger, K. E.; Tuckerman, M. E., Ab Initio Molecular Dynamics Simulation of the Structure and Proton Transport Dynamics of Methanol-Water Solutions. *Journal of Physical Chemistry B* **2006,** 110, 3712-3720.

(12)Gold, S.; Chu, K.-L.; Lu, C.; Shannon, M. A.; Masel, R. I., Acid Loaded Porous Silicon as a Proton Exchange Membrane for Micro-Fuel Cells. *Journal of Power Sources* **2004,** 135, (1-2), 198-203.

Prediction of Multicomponent Diffusion Coefficients in Liquids

Sebastian Rehfeldt and Johann Stichlmair, Technische Universitaet Muenchen, Garching, Germany

Abstract

Knowledge of diffusion coefficients in liquids is of great interest for the calculation and simulation of mass transfer processes. In the literature, several prediction models for binary diffusion coefficients are found. For multicomponent systems, however, only a few models exist. Due to lack of ternary diffusion data, these models have not been verified for real systems until now. To overcome this limitation, multicomponent diffusivities were measured within the whole concentration space of several ternary systems. Fick diffusivities were transformed to less concentration dependent Maxwell-Stefan diffusivities using the thermodynamic correction factor. Four prediction models were tested by a comparison of predicted values to experimental data. In some systems, multicomponent diffusion prediction shows promising results. However, the quality of predicted diffusion coefficients strongly depends on an accurate thermodynamic description of the system.

Introduction

The physical phenomenon of diffusion is omnipresent in every natural as well industrial process involving mass transfer. In many cases, diffusion plays an important role as the rate limiting mechanism. Therefore, the determination of diffusion coefficients in liquids is of great interest for the calculation and simulation of mass transfer processes. Prediction of diffusion coefficients can be divided in several steps. The first step is the determination of the diffusion coefficient at infinite dilution. These coefficients are the initial points for determination of diffusion coefficients in concentrated mixtures. Most studies about diffusion concentrate on measuring and predicting diffusivities in binary mixtures. Research on this topic advanced, also due to a fair database of experimentally determined binary diffusivities. However, real processes mostly deal with multicomponent mixtures. In this case, both the experimental and theoretical investigations of diffusion are much more complex. Therefore, few multicomponent diffusion data are found in the literature.

Theory

There are two well established theories describing diffusion. The more commonly used is Fick's Law [1]:

$$j_i = -\sum_{j=1}^{n-1} D_{ij} \cdot \nabla c_j \qquad i = 1 \ldots n-1 \qquad (1)$$

The flow j is defined as the product of diffusion coefficients D_{ij} and a driving force, i.e. the concentration gradient ∇c_j.

In case of a binary mixture, only one diffusion coefficient is needed to describe all diffusional fluxes. However, in systems with more than two components, more coefficients are necessary and the structure of the flux equations is getting more complex. For example, four diffusivities D_{ij} are required to describe diffusion in a ternary system:

$$\begin{bmatrix} j_1 \\ j_2 \end{bmatrix} = -\begin{bmatrix} D_{11} & D_{12} \\ D_{21} & D_{22} \end{bmatrix} \cdot \begin{bmatrix} \nabla c_1 \\ \nabla c_2 \end{bmatrix} \tag{2}$$

The coefficients D_{11} and D_{22} are called the main coefficients. They show the dependency of the flow rate of a component on its own concentration gradient. The influence on the flow rate of the concentration gradient of the other component is taken into account by the cross coefficients D_{12} and D_{21}. However, only $n-1$ flow rates are independent. In the ternary case, the third component is called the "solvent", its flow rate can be determined with the flux balance condition. The four ternary diffusivities implicitly contain information on the flux of the solvent, so their values depend on its choice. It is obvious that ternary Fick diffusivities are not related to the binary diffusion coefficients. This is a great disadvantage of Fick's Law.

Maxwell and Stefan developed a physically more consistent diffusion theory [2]. Based on the kinetic theory of ideal gases, the assumption of the Maxwell-Stefan-equations is an equilibrium of molecular friction and thermodynamic interaction. Any deviation from equilibrium leads to a diffusion flux. For liquid, non-electrolytic mixtures, the chemical potential gradient $\nabla\mu_i$ is used as the driving force d, which is counterbalanced by friction. The friction between two components is proportional to their velocity difference $v_i - v_j$ and their amount of molar fractions x:

$$d_i = \frac{x_i}{R \cdot T} \cdot \nabla_{(T,p)}\mu_i = \sum_{j=1}^{n-1} \Gamma_{ij} \cdot \nabla x_j = -\sum_{j=1}^{n} \frac{x_i x_j}{Đ_{ij}} \left(v_i - v_j\right) \tag{3}$$

Here, $Đ_{ij}$ denote the Maxwell-Stefan diffusivities which can be interpreted as inverse drag coefficients of components i and j. The thermodynamic correction factor Γ_{ij} is calculated from the activity coefficients γ_i:

$$\Gamma_{ij} = \delta_{ij} + x_i \left.\frac{\partial \ln \gamma_i}{\partial x_j}\right|_{T,p} \quad \text{with} \quad \delta_{ij} = 1 \text{ for } i = j \qquad \text{and } \delta_{ij} = 0 \text{ for } i \neq j \tag{4}$$

In binary systems, there again is a single diffusion coefficient $Đ_{ij}$. In contrast to Fick's law, for description of diffusion in a ternary mixture only three Maxwell-Stefan diffusivities $Đ_{ij}$ are needed. Friction of each couple of components is taken into account by one coefficient. Thus, the choice of a solvent is dispensable. Therefore, multicomponent Maxwell-Stefan diffusivities can be related to the binary case. This should be advantageous for prediction.

In the literature, most publications only contain Fick diffusivities, since they can be measured experimentally. Binary Maxwell-Stefan diffusivities can be calculated from experimental Fick diffusion coefficients and the thermodynamic correction factor:

$$Đ = D/\Gamma \tag{5}$$

Transformation is fairly simple in binary systems. Regarding ternary systems, the situation is more complex. Indeed, there is an algebraic relationship between Fick and Maxwell-Stefan diffusion coefficients [1]. Since there are four Fick coefficients, but only three Maxwell-Stefan coefficients, the transformation is only straightforward from Maxwell-Stefan to Fick. Transforming Fick diffusivities to Maxwell-Stefan, the system of equations is over-determined. Fick diffusion coefficients and thermodynamic factors both have experimental errors and are, in turn, usually inconsistent [3]. Therefore, a least-square fitting is required to compute Maxwell-Stefan diffusivities from Fick diffusivities.

Prediction Models

In the literature, a large number of prediction models can be found for diffusion coefficients in concentrated binary mixtures [1]. Initial points for those calculations are the diffusion coefficients at infinite dilution. These coefficients can be calculated from several models found in the literature (e.g. [1], [4]), or be determined by experiments. Since those prediction models show wide error ranges [4], experimentally determined diffusion coefficients at infinite dilution are recommended for prediction in concentrated mixtures.

Early prediction models for diffusion coefficients in concentrated binary liquids have been developed for Fick diffusion coefficients. Since most models contain some kind of thermodynamic correction factor similar to Γ, those models have been transformed to yield the Maxwell-Stefan diffusion coefficient. This result can be transformed with Eq. 5 for prediction of the Fick diffusion coefficient.

One basic prediction model originates from a proposal by Darken [5]. Here, the binary diffusion coefficient is calculated as a sum of the binary diffusivities at infinite dilution multiplied with the molar fraction:

$$Đ_{12} = D_{12}^{x_1 \to 1} \cdot x_1 + D_{12}^{x_1 \to 0} \cdot x_2 \tag{6}$$

Later, Vignes [6] suggested this equation for concentrated binary liquids:

$$Đ_{12} = \left(D_{12}^{x_1 \to 1}\right)^{x_1} \cdot \left(D_{12}^{x_1 \to 0}\right)^{x_2} \tag{7}$$

A third model is an extension of the Vignes-equation by a viscosity correction, which was first porposed by Leffler and Cullinan [7]:

$$Đ_{12} = \left[\left(D_{12}^{x_1 \to 1} \cdot \eta_1\right)^{x_1} \cdot \left(D_{12}^{x_1 \to 0} \cdot \eta_2\right)^{x_2}\right] / \eta \tag{8}$$

Here, η_i denotes the viscosity of a single species, η is the viscosity of the mixture. In the literature, several other modifications of the Vignes-equation are found (e.g. [8]) and usually show better results for certain mixtures. Nevertheless, the Vignes-equation is recommended in general [1], [4] since it is well tested and easy to use.

Prediction methods for ternary Maxwell-Stefan diffusivities start with a ternary extension of the binary Vignes-equation [9]:

$$Đ_{ij} = \left(Đ_{ij}^{x_j \to 1}\right)^{x_j} \cdot \left(Đ_{ij}^{x_i \to 1}\right)^{x_i} \cdot \left(Đ_{ij}^{x_k \to 1}\right)^{x_k} \tag{9}$$

The first two diffusion coefficients at infinite dilution can be derived from binary experiments. Values for the third diffusion coefficient at infinite dilution are unknown. Therefore, several authors tried to find an expression for this unknown coefficient. Wesselingh and Krishna developed this equation [9]:

$$Đ_{ij} = \left(Đ_{ij}^{x_j \to 1}\right)^{x_j} \cdot \left(Đ_{ij}^{x_i \to 1}\right)^{x_i} \cdot \left(Đ_{ij}^{x_j \to 1} \cdot Đ_{ij}^{x_i \to 1}\right)^{x_k/2} \tag{10}$$

Here, $Đ_{ij}$ at infinite dilution in component k is replaced by the geometric average of the other two diffusion coefficients at infinite dilution. Using this model, the ternary diffusion coefficient $Đ_{ij}$ is completely independent of the nature of the third component k, which seems physically doubtful.

Kooijman and Taylor use the geometric average of other diffusion coefficients at infinite dilution in their prediction model [10]:

$$Đ_{ij} = \left(Đ_{ij}^{x_j \to 1}\right)^{x_j} \cdot \left(Đ_{ij}^{x_i \to 1}\right)^{x_i} \cdot \left(Đ_{ik}^{x_k \to 1} \cdot Đ_{jk}^{x_k \to 1}\right)^{x_k/2} \tag{11}$$

In this model, the ternary diffusion coefficient $Đ_{ij}$ is also a function of the binary diffusivities $Đ_{ik}$ and $Đ_{jk}$.

Rutten extended the model of Wesslingh and Krishna with a viscosity correction [11]. By this way, component k has an influence on the ternary diffusivity $Đ_{ij}$.

$$Đ_{ij} = \left(Đ_{ij}^{x_j \to 1}\right)^{x_j} \cdot \left(Đ_{ij}^{x_i \to 1}\right)^{x_i} \cdot \left(Đ_{ij}^{x_j \to 1} \cdot Đ_{ij}^{x_i \to 1} \cdot \frac{\sqrt{\eta_i \cdot \eta_j}}{\eta_k}\right)^{x_k/2} \tag{12}$$

Additionally, a fourth model is proposed by Rehfeldt and Stichlmair [12]:

$$Đ_{ij} = \left(Đ_{ij}^{x_j \to 1}\right)^{x_j} \cdot \left(Đ_{ij}^{x_i \to 1}\right)^{x_i} \cdot \left(Đ_{ij}^{x_j \to 1} \cdot Đ_{ij}^{x_i \to 1} \cdot Đ_{ik}^{x_k \to 1} \cdot Đ_{jk}^{x_k \to 1}\right)^{x_k/4} \tag{13}$$

Using the geometric average of all four diffusion coefficients at infinite dilution of the models developed by Wesselingh and Krishna and Kooijman and Taylor, this model constitutes a combination of those approaches. The diffusion coefficient $Đ_{ij}$ depends on component k, but comparing to the model of Kooijman and Taylor, the influence of the foreign binary diffusivities $Đ_{ik}$ and $Đ_{jk}$ is reduced.

Experimental Setup

Fick diffusion coefficients were experimentally determined by holographic laser-interferometry [13]. In this method, a measuring cell is filled with two mixtures with slightly different concentrations. Unsteady diffusion causes a changing concentration profile inside the diffusion cell, which causes an interference pattern. The cell is photographed in intervals with a digital camera. Analyzing the change of the interference pattern, Fick diffusion coefficients can be calculated [14]. For binary datapoints, only a single experiment is necessary. Determination of ternary Fick diffusivities is much more complex. Several experiments with pairs of mixtures with different concentrations have to be performed.

Binary Results

Fick diffusion coefficients were measured over the whole concentration range of several binary and ternary systems. Subsequently, Maxwell-Stefan-diffusivities were computed from the experimental Fick diffusivities employing the Wilson excess Gibbs energy model for calculation of the thermodynamic correction factor [1]. Results of binary diffusion experiments in the ternary system acetone – 1-butanol – 1-propanol show good agreement with binary prediction models.

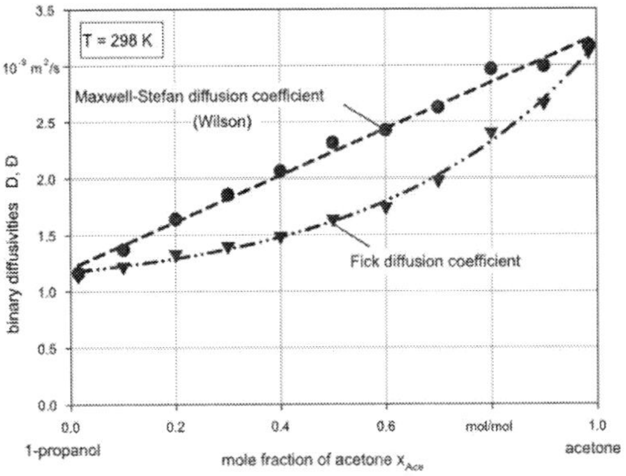

Fig. 1: Binary diffusivities acetone – 1-propanol.

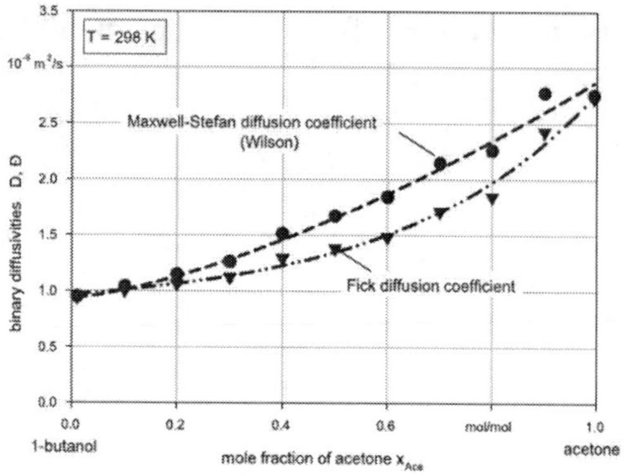

Fig. 2: Binary diffusivities acetone – 1-butanol.

The binary Fick diffusion coeffcicents of acetone – 1-propanol (Fig. 1) and acetone – 1-butanol (Fig. 2) show a similar dependency on molar fraction of acetone. Values in the binary system acetone – 1-propanol are slightly higher. The Maxwell-Stefan diffusion coefficients in both systems are less dependent on concentration. In acetone – 1-propanol, a linear relation of the Maxwell-Stefan diffusivity and concentration can be observed, as predicted by Darken (Eq. 6). The graph of the Maxwell-Stefan diffusion coefficient in the binary system acetone – 1-butanol is slightly curved, a result which resembles the prediction model of Vignes (Eq. 7).

The system 1-propanol – 1-butanol (not shown) is virtually a thermodynamic ideal mixture. The Fick diffusion coefficient has roughly constant values. Due to values of the thermodynamic correction factor near unity, there is only a small difference between Fick and Maxwell-Stefan diffusivities.

The experimental results shown prove that in these binary mixtures, Maxwell-Stefan diffusivities can be predicted quite accurately, if diffusivities at infinite dilution are known and an exact thermodynamic description of the system is at hand. However, errors in the thermodynamic phase equilibrium data have a big influence on the quality of the Maxwell-Stefan diffusion coefficients. Since the thermodynamic factor involves the first derivative of the activity coefficient with respect to composition, MS coefficients are quite sensitive to the parameters used for calculating the thermodynamic correction factor. These parameters are fitted against vapor-liquid equilibrium data. Even if the fitted parameters lead to good results for phase equilibrium, the first derivatives of the activity coefficients can show large deviations [1], [3].

By substituting water for 1-butanol in this ternary system, the non-ideal character increases and, in term, the thermodynamic description of the binary systems is more complicated.

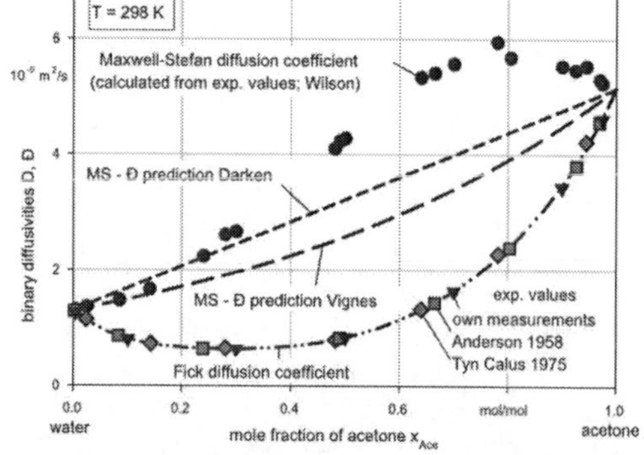

Fig. 3: Binary diffusivities acetone – water.

In Fig. 3, experimental Fick diffusivities of different authors ([15], [16]) and own measurements are displayed. All measured points combine to a clean run of the curve of the Fick diffusion coefficient. Thermodynamic non-idealities lead to a minimum of the Fick diffusion coefficient at a mole fraction of acetone of about 0.3. Experimental data points are again transformed into Maxwell-Stefan diffusion coefficients by applying Eq. 5 with usage of the thermodynamic correction factor calculated with Wilson. Here, the Maxwell-Stefan diffusivities deviate from both Darken and Vignes prediction models, especially at higher concentrations of acetone.

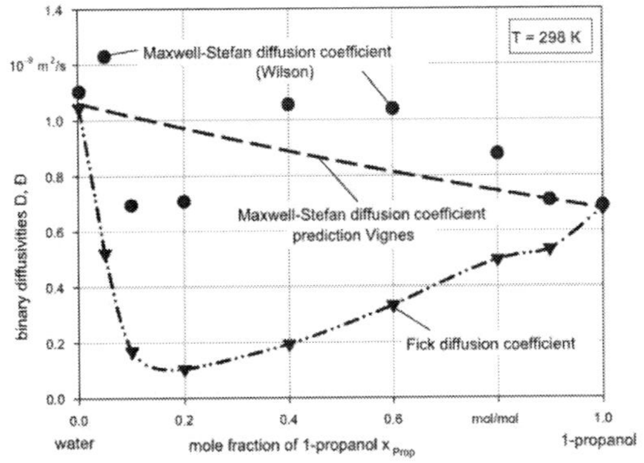

Fig. 4: Binary diffusivities 1-propanol – water.

A similar result is observed in the binary system 1-propanol – water (Fig. 4). Again, the Fick diffusion coefficient runs through a minimum. No prediction model can correctly reflect the Maxwell-Stefan diffusion coefficients. In regard to the prediction model of Vignes, some points are well below, others are above the run of curve. At least, the Maxwell-Stefan diffusion coefficient shows a less concentration dependent behavior than the Fick diffusion coefficient.

Ternary Results

Ternary Fick diffusion coefficients were transformed into ternary Maxwell-Stefan diffusion coefficients [1]. Contrary to the ternary Fick diffusivities, ternary Maxwell-Stefan diffusivities can be related to the relevant binary diffusivities. Regarding the binary edges of the ternary diffusion coefficients of acetone – 1-butanol – 1-propanol (Fig. 5), the shapes of the diffusion coefficient surfaces

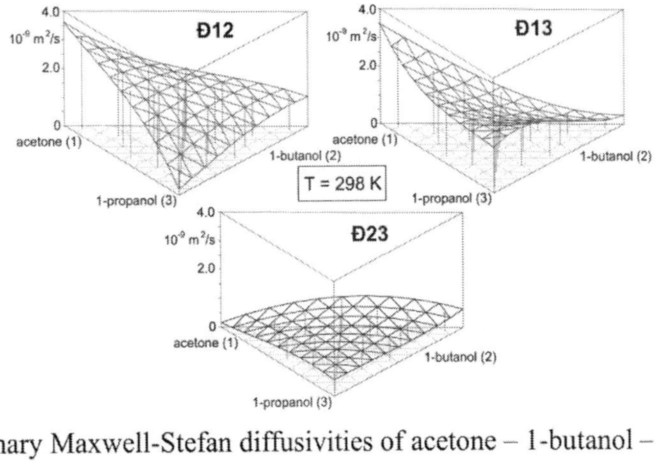

Fig. 5: Ternary Maxwell-Stefan diffusivities of acetone – 1-butanol – 1-propanol.

resemble their respective binary coefficients. Noteworthy, only ternary diffusion coefficients were used to fit the surface. Prediction of ternary Maxwell-Stefan diffusion coefficients usually starts with values of the binary diffusion coefficients at the binary edge. Therefore, behavior of the diffusion coefficient with increasing molar fraction of the respective third component is interesting. Here, values of all three Maxwell-Stefan diffusivities decrease while approaching the opposite corner of the triangle.

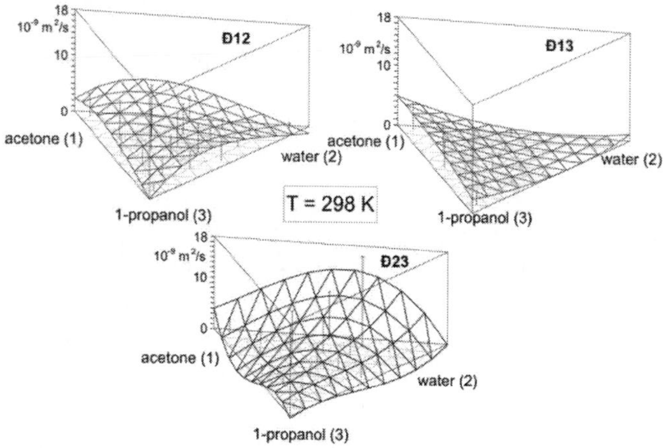

Fig. 6: Ternary Maxwell-Stefan diffusivities of acetone – water – 1-propanol.

The ternary Maxwell-Stefan diffusion coefficients of acetone – water – 1-propanol (Fig. 6) show a different image. While the Maxwell-Stefan diffusivities of acetone – 1-butanol – 1-propanol exhibited a describable concentration dependence as expected, especially the ternary coefficients *Đ12* and *Đ23* of acetone (1) – water (2) – 1-propanol (3) vary significantly with concentration and take rather high values compared to predicted ones (Fig. 7, predicted with Eq. 13).

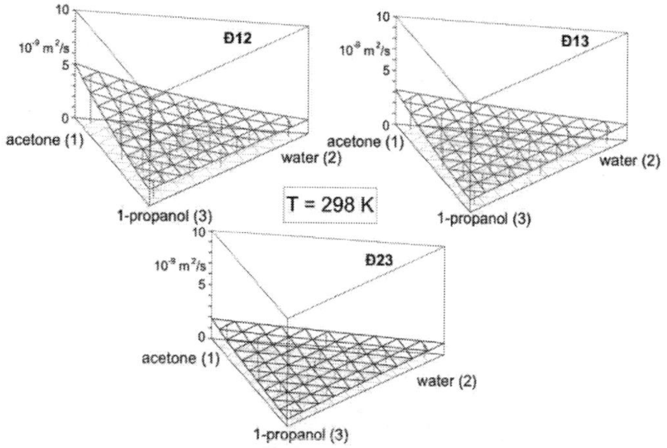

Fig. 7: Predicted (Eq. 13) Maxwell-Stefan diffusivities of acetone – water – 1-propanol.

Noteworthy, the Maxwell-Stefan diffusivities of the respective binary systems (Fig. 3, Fig. 4) also showed higher values than predicted. Since the thermodynamic correction factor used for transformation to Maxwell-Stefan diffusion coefficients is calculated from the same Wilson parameters as for binary transformation, this behavior is not surprising. Obviously, ternary transformation is more sensitive to inaccuracies in thermodynamic data, because deviations from prediction are higher than in the binary case.

Discussion

Regarding the results of the ternary system acetone – 1-butanol – 1-propanol, the model of Rutten shows the largest deviations from experimental data (Fig. 8). The viscosity correction has obviously too much weight here. Although some points can be predicted quite accurate, the model of Kooijman and Taylor predicts several values with deviations of more than 50%. The model of Wesselingh and Krishna works quite well for this ternary system. Questionable is, of course, the independency of the diffusion coefficient $Đ_{ij}$ on component k. This model predicts the same results for the ternary diffusion coefficient $Đ13$ of acetone and 1-propanol in this system as well as in acetone – water – 1-propanol. The best results for this ternary system are delivered by the approach of Rehfeldt and Stichlmair. The average deviation is 23.9%, which seems quite good for prediction of ternary diffusivities in the whole concentration range of a system. The results of prediction of the ternary system of acetone – water – 1-propanol (Fig. 9) are less satisfactory.

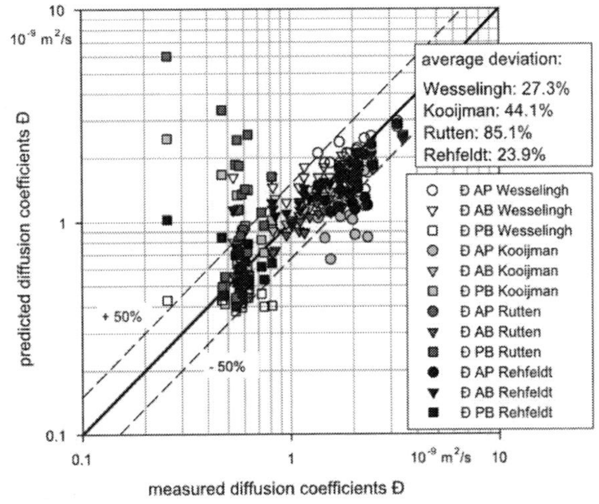

Fig. 8: Comparison of several prediction methods for acetone – 1-butanol – 1-propanol.

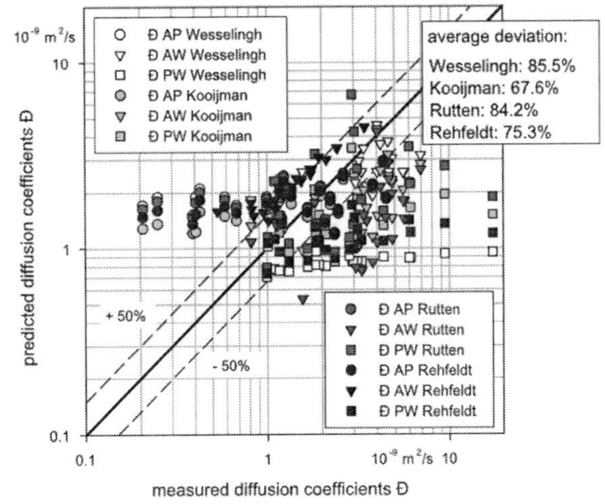

Fig. 9: Comparison of several prediction methods for acetone – water – 1-propanol.

All prediction models show much larger deviations. Contrary to acetone – 1-butanol – 1-propanol, there are nearly no points close to the diagonal of the error diagram in Fig. 9, most of them are spread over the whole plot. Measured diffusion coefficients range from 0.2×10^{-9} m²/s to 17×10^{-9} m²/s, while predicted diffusion coefficients only range from 0.5×10^{-9} m²/s to 7×10^{-9} m²/s. Obviously, inaccuracies in thermodynamic data, as already seen in the binary systems, complicates prediction for this ternary system.

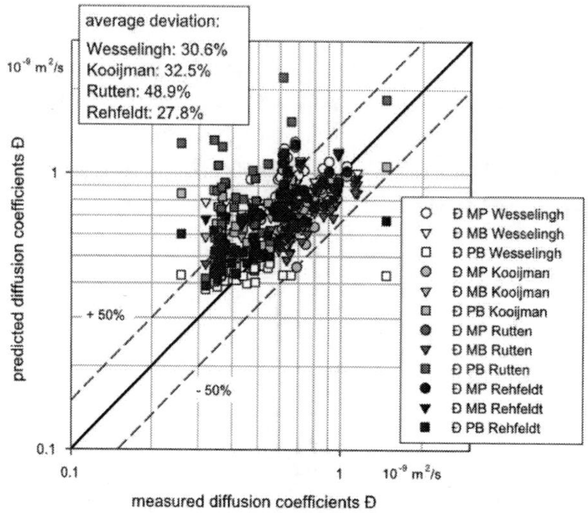

Fig. 10: Comparison of several prediction methods for methanol – 1-butanol – 1-propanol.

Additionally, results of prediction in another ternary system, methanol – 1-butanol – 1-propanol are presented in Fig. 10. Since all three substances are univalent alcohols, this system has only slight non-idealities. Again, the model of Rehfeldt and Stichlmair delivers the best result with an average deviation of 27.8%, although the models of Wesselingh and Krishna and of Kooijman and Taylor produce results of quite similar quality. For this ternary system, the model of Rutten shows the largest inaccuracies.

Conclusions

Binary and ternary Fick diffusion coefficients were measured by holographic interferometry in several systems. Maxwell-Stefan diffusion coefficients were derived from the experimental Fick diffusivities using the thermodynamic correction factor. Binary VLE-data was used with the Wilson excess-enthalpy model for calculation of the thermodynamic correction factor. Several ternary prediction methods were tested against the experimental data.

Regarding the results of all three ternary systems investigated, it is obvious, that the success of prediction depends on an accurate thermodynamic description of the system. Systems with low non-idealities, (methanol – 1-butanol – 1-propanol and acetone – 1-butanol – 1-propanol) yield better results on prediction than the strongly non-ideal system acetone – water – 1-propanol. This is already indicated by the results of the binary sub-systems, where large deviations from the prediction models of Darken and Vignes can be found in the diffusivities of acetone – water and 1-propanol – water.

The model of Rehfeldt and Stichlmair, which combines the approaches of Wesslingh and Krishna and Kooijman and Taylor yields the best results for the ternary systems considered, but has also high inaccuracies in the system acetone – water – 1-propanol. For the systems with less non-idealities, the models of Wesslingh and Krishna and of Kooijman and Taylor deliver quite good results. The model of Rutten cannot be recommended for the systems investigated here.

A general evaluation of ternary prediction models, it is reasonable to use systems with accurate thermodynamic data. Since most real processes deal with non-ideal mixtures, it is essential for application of multicomponent diffusion prediction to improve thermodynamic description of those systems.

References

[1] R. Taylor, R. Krishna, Multicomponent mass transfer, Wiley&Sons Inc., New York 1993.

[2] J. Stefan, Sitzungsber. Akad. Wiss. Wien, 63, Abt. II (1871), 63-124.

[3] A. Wild, Multicomponent Diffusion in Liquids, Ph.D. thesis, TU Munich, VDI-Verlag, Düsseldorf, 2003.

[4] B.E. Poling, J.M. Prausnitz, J.P. O'Connell, The Properties of Gases and Liquids, McGraw-Hill, New York, 2001.

[5] L.S. Darken, Am. Inst. Min. Metall. Pet. Eng., 175 (1948), 184-201.

[6] A. Vignes, IEC Fund., 5 (1966) 2, 189-199.

[7] J. Leffler, H.T. Cullinan, Ind. Eng. Chem. Fundam 9 (1) (1970) 84-88.

[8] D. Bosse, H.J. Bart, Ind. Eng. Chem. Res. 45 (5) (2006) 1822-1828.

[9] J.A. Wesselingh, R. Krishna, Mass Transfer, Ellis Horwood, Chichester, England, 1990.

[10] H.A. Kooijman, R. Taylor, Ind. Eng. Chem. Res., 30 (6) (1991), 1217-1222.

[11] Ph.W.M. Rutten, Diffusion in Liquids, Delft University Press, 1992.

[12] S. Rehfeldt, J. Stichlmair, Fluid Phase Equilib. (2006), doi:10.1016/j.fluid.2006.10.008

[13] M. Pertler, Die Mehrkomponenten-Diffusion in nicht vollständig mischbaren Flüssigkeiten, Ph.D. thesis, TU Munich, 1996.

[14] D.G. Miller, J. Phys. Chem., 90 (1988), 1706-1717.

[15] D.K. Anderson, J.R. Hall, A.L. Babb, J. Phys. Chem. 62 (4) (1958) 404-409.

[16] M.T. Tyn, W.F. Calus, J. Chem. Eng. Data 20 (3) (1975) 310-316.

Estimation of Activity Coefficients in Binary Systems Using Neural Networks

L. Antonio Estévez
University of Puerto Rico
Department of Chemical Engineering
Mayagüez PR 00681-9046

Harry Rodríguez-Vallés
Department of Quality Assurance
Cordis LLC a Johnson & Johnson Company
PO Box 5125
San Germán, PR 00683

Nazario D. Ramirez-Beltran
University of Puerto Rico
Department of Industrial Engineering
Mayagüez PR 00681-9043

ABSTRACT

About 13,000 chemical interactions among functional groups have been identified in chemical products and those interactions were organized into the information matrix. A feed-forward neural network was trained with a large set of phase-equilibrium data and the functional groups from chemical compounds were used to predict the interaction coefficients. Once interactions constants were known, a chemical thermodynamic model was used to predict the activity coefficients. The proposed estimation procedure is especially useful for those cases when phase-equilibrium data for a given system are not available. The structural formula of a given chemical product is decomposed into a set of functional groups. The interactions formed among the involved functional groups are selected from the information matrix to create the training patterns for a given mixture, which include the molecular weights and the number of units of the functional groups. The minimum variance criterion is proposed to identify the appropriate structure of a feed-forward neural network. Five chemical mixtures were arbitrarily selected from DECHEMA data. Neural networks were identified and the activity coefficients were successfully estimated using the proposed methodology.

INTRODUCTION

A neural network approach is used to estimate the activity coefficients of chemical compounds in liquid solutions. The activity coefficients are crucial in the design of separation processes such as distillation and absorption. The main purpose of this research is to reduce the need of experimental equilibrium data that are usually costly and time consuming. Usually, phase-equilibrium experiments are needed to determine the activity coefficients (Smith et al., 1996). The key of this research is that a chemical compound is formed by different functional groups. Thus, a chemical compound can be built by combining the right functional groups, and as a result, interaction between functional groups can be grouped to estimate the interactions among chemical compounds of a new chemical product.

The group contribution method has been shown to be effective to predict the activity coefficients of various chemical mixtures. This method has been modified and extended to many more groups by Constantinous and Gani (1994). They proposed a new group contribution technique for estimating important physical and thermodynamic properties of pure compounds. They use the molecular structure of a compound so that significantly improved prediction of properties can be accomplished. They proposed a method where the molecular structure of a compound is considered to be a collection of two types of groups: simple functional groups (or first-order groups), and functional groups formed by the first-order groups as building blocks (or second-order groups). The second-order groups are based on the principle of conjugation. Once the group is identified, an empirical model was fitted to the given data to estimate some parameters that allow estimating physical properties of pure components with high accuracy. Essentially, they proposed a general linear regression model that relates the contribution of the first-order and the second-order groups to several linear and nonlinear response variables. The response variable is associated to a particular property, for instance, critical temperature, critical pressure, critical volume, melting point, normal boiling point, standard Gibbs free energy, and standard enthalpy of vaporization and formation. They estimated a set of parameters, which are included in mathematical functions, to express the physical and thermodynamic properties of pure compounds. The Constantinou and Gani method was designed under the assumption that the response variable can be expressed as a linear combination of the group contributions. However, if the appropriate relation is nonlinear, then the nonlinear regression model can be obtained by trial an error.

In this research the artificial neural network methodology is proposed to model the linear or nonlinear relationships among the functional groups and equilibrium data. Using the concept of functional groups contribution, a chemical compound can be broken down into its groups. A molecule of a chemical compound is divided up into smaller pieces that are known as functional groups, which are characterized by specific molecular structures and physical properties (Fredenslund et al., 1975). Thus, the functional groups can be brought together to makeup a molecule in a mixture for which no equilibrium data is available.

A large library of functional groups has been developed and a feedforward neural network has been trained to estimate the interactions among functional groups. Thus, it is possible to combine the interactions of the functional groups in a mixture, and reliable estimates of activity coefficients can be obtained under the framework of the artificial neural networks. Five mixtures were arbitrarily selected from a large library of equilibrium data to test the proposed approach and successful results were obtained.

A careful attention should be devoted to the application of neural networks, since an automatic application of the neural network may lead to misleading predictions. A second important contribution of this research is a proposed systematic method to identify the neural network structure. *The minimum-variance estimation criterion* is proposed in this paper to ensure that the identification of a network structure provides a reliable prediction scheme. The accuracy of the prediction tool is illustrated by comparing the neural network approach to the conventional estimation procedures whose estimation is based on phase-equilibrium data.

This paper is organized as follows. The second section presents a brief definition of the activity coefficient. The third section describes the chemical thermodynamic models used to estimate the activity coefficients. Section fourth presents the proposed neural network approach to estimate the activity coefficients based on a library of equilibrium data; this section also presents a proposed methodology to identify the appropriate neural network structure for a given input output process. The fifth section illustrates a step-by-step example of the proposed estimation procedure, and the last section presents some conclusions.

THERMODYNAMIC MODELS

The activity coefficients of individual components in a mixture are essential to model separation processes. Estimation of the activity coefficient requires selecting a thermodynamic model and estimating the model constants. Regression approach and phase-equilibrium data are used to estimate the model constants. Among the best known thermodynamic models are the following: Margules, van Laar, Wilson, Non-Random Two Liquids (NRTL), and UNIversal QUAsi-Chemical (UNIQUAC) models (Abrams, and Prausnitz, 1975; Smith, el al., 1996). These methods express the activity coefficient as a function of the mole fraction and model constants. In this research the NRTL method was selected to estimate the activity coefficient because it is a local-composition model and it requires a small amount of parameters as well as it provides reliable estimations. The NRTL model for multicomponent mixtures can be written as follows:

$$\ln\gamma_i = \frac{\sum_{j=1}^{c}\tau_{ij}G_{ji}x_j}{\sum_{h=1}^{c}G_{hi}x_h} + \sum_{j=1}^{c}\left(\frac{x_j G_{ij}}{\sum_{h=1}^{c}G_{hj}x_h}\left[\tau_{ij} - \frac{\sum_{k=1}^{c}\tau_{kj}G_{kj}x_k}{\sum_{h=1}^{c}G_{hj}x_h}\right]\right) \quad \textbf{(1)}$$

where:

$$A_{ij} = T\tau_{ij} \quad \textbf{and} \quad G_{ij} = \exp(-\alpha_{ij}\tau_{ij}) \quad \textbf{(2)}$$

A_{ij} are the interaction constants between the i and j compounds, α_{ij} is the NRTL constant between the i and j compounds, x_i is the mole fraction for the ith compound, and T is temperature. Therefore, the estimation of the activity coefficient problem reduces to estimate A_{ij} and α_{ij}. The proposed methodology to estimate the activity coefficient is developed under the framework of neural networks and contribution of functional group. It should be pointed out that NRTL constants were used to train the neural network, i.e., the actual phase-equilibrium data were not considered to train the neural network to reduce the input output in the neural network.

NEURAL NETWORK ESTIMATION PROCEDURE

The conventional approach to obtain the individual activity coefficients in a mixture requires experimental equilibrium data. However, given that experiments are expensive and time consuming, the main purpose of this research is to be able to estimate the activity coefficients without experimental data. This approach is possible because there is a large equilibrium database that can be used to train a neural network to recognize the interaction among functional groups (Sorensen and Arlt, 1979).

The entertained methodology in this research is based on dividing up each component of a mixture into its constituents functional groups, identify the required interactions among functional groups, and then train a neural network to translate functional-group interactions into component interaction. Thus, once the neural network is trained, the interaction constants for products where no equilibrium data are available will be estimated.

An extensive chemistry data series (Sorensen and Arlt, 1979) was the source of phase-equilibrium data and the corresponding NRTL constants. About 13,000 chemical interactions among functional groups have been identified as the most useful ones and they are organized into the information matrix of approximately 5,000 by 13,000 to train a neural network. Each element of the information matrix, $X_{h,k}$, is a functional-group coefficient and will be located in the n_{hk} column. Each column of the information matrix represents a single chemical interaction between the h and k functional groups. Each row of the information matrix represents a chemical compound for which experimental equilibrium data are available. Thus, the location of the column associated to a given interaction was computed by using the following formula:

$$n_{h,k} = g(h-1)+k \qquad \textbf{(3)}$$

It should be noted that $n_{h,k}$ is the column of the information matrix that contains the chemical interactions corresponding to the h and k functional groups, g is the total number of functional groups, h and k are the numbers associated to the first and the second functional groups, respectively. For instance, suppose that there are five functional groups and we want to compute the interaction between the second and fourth functional groups, the corresponding columns in the information matrix are computed as follows:

$$n_{2,4} = 5(2-1)+4 = 9 \qquad \textbf{and} \qquad n_{4,2} = 5(4-1)+2 = 17$$

Thus, the coefficient $X_{2,4}$ will be located in column 9, and the coefficient $X_{4,2}$ will be located in column 17. The functional-group coefficient is computed by using the following expression:

$$X_{hk} = a_h a_k \frac{w_h}{w_k} \qquad \textbf{(4)}$$

where a_h is the amount of units interacting of the functional group h, w_h is molecular weight of the h functional group, a_k and w_k are defined in a similar manner as the previous definitions.

One of the main purposes of this research is to derive an estimate of the interaction between two compounds, knowing that each compound has several functional groups. Thus, a possible estimator of the interaction constant between compounds i and j could be a linear combination of the contribution of each functional-group interaction. That is,

$$A_{i,j} = \sum_{\{h,k\}\subset H} \sum_{\{k,h\}\subset K} \beta_{h,k} X_{h,k} \qquad \textbf{(5)}$$

Where $\beta_{h,k}$ are parameters that are estimated from equilibrium experimental data and A_{ij} is the interaction constant obtained from phase equilibrium data. H is a set of all subscripts that contain the interactions between h and k, and K is a set of all subscripts that contain the interactions between k and h. It should be noted that the interaction between h and k is different from k and h. Since the NRTL constants (A_{ij} and A_{ji}) are not symmetric, i.e., the first functional group is the actuator and the second one is the receiver and knowing that the strength of each functional group is different. The previous model can be improved after converting it into a nonlinear model, i.e., the interaction constants between compounds i and j is represented by the following model:

$$A_{i,j} = \sum_{\{h,k\}\subset H} \quad \sum_{\{k,h\}\subset K} \beta_{h,k} X_{h,k}^{m_{h,k}} \qquad (6)$$

where $m_{h,k}$ are the exponents to be estimated from data for improving the model fitting process. The major difficulty in fitting a nonlinear model is to derive the appropriate starting point that after some iterations the algorithm eventually arrives to the optimal solution. An efficient procedure to perform nonlinear modeling and suggested in this research is to use the linear estimation as the initial point for the iterative procedure. A nonlinear optimization technique is then selected to find the estimators that minimize the difference between the observed and the estimated values. Successful applications of the Marquardt algorithm have been reported values (Ramirez-Beltran et al., 1998; Bazaraa, et al., 1993). The Marquardt algorithm takes into account the advantages of the steepest descent and the Newton methods to improve the convergence process.

A neural network modeling technique is proposed to increase prediction accuracy. The neural network model technique can be viewed as a general nonlinear modeling approach. The training patterns of the neural networks are formed by the input and output vectors. The X's form the elements of the input vector, P, and the corresponding interaction constant, A_{ij}, forms the output target, T. That is,

$$P' = \begin{bmatrix} X_{h_1,k_1} & X_{k_1,h_1} & \cdots & X_{k_s,h_s} \end{bmatrix} \quad \textbf{and} \quad T = \begin{bmatrix} A_{i,j} \end{bmatrix} \qquad (7)$$

where the selected subscripts h,k and k,h belong to the set of H and K, respectively, and s is the total number of functional-group interactions identified in the information matrix for a given compound.

A neural network with two layers was selected. Neural networks with three layers or more were not considered since in this case they overparameterize the learning process and prediction capability. A neural network composed with a single layer was not used because it does not have the capability of modeling nonlinear relationships, and the activity coefficient is characterized by having a highly nonlinear relationship with mole fraction, temperature, and NRTL constants. Thus, the identification problem reduces to determine the number of neurons in the hidden layer. Therefore, the suggested identification procedure for determining the neural network structure consists of selecting a single neuron in the hidden layer and then observe the fitting accuracy by means of the sum of square error (sse), and at the end of the training task an estimator of A_{ij} is recorded. This process is repeated at least five times, and the sample variance of the estimator is computed. That is,

$$\mathrm{var}(\hat{A}_{ij}) = \frac{1}{m-1}\sum_{i=1}^{m}\left(\hat{A}_{ij} - \overline{A}\right)^2 \qquad \textbf{(8)}$$

where $\hat{A}_{ij}$ is the network estimator of A_{ij}, $\overline{A}$ is the average of $\hat{A}_{ij}$'s, and m is the number of times that A_{ij} was estimated by the neural network for a given structure.

The number of neurons in the hidden layer is increased by one until the $\min\{\mathrm{var}(\hat{A}_{ij})\}$ is identified in the selected structure. The described process is repeated with different structures. Finally, the best neural network structure is the one that provides *the minimum variance estimator with the least number of neurons*. Thus, if the network provides an unbiased estimator we will obtain the minimum variance unbiased estimator. On the other hand, if the network exhibits a biased estimator then the bias will be small since we are selecting the neural network with the minimum number of neurons, i.e., the principle of parsimony will help in reducing the bias (Box and Jenkins, 1976; Wei, 1990; Pandi and Wu, 1983). It should be noted that each replicate is computed with different initial weight matrices with the purpose of identifying the minimum variance estimator for a given network structure. The structures in this context are defined according to the transfer function included in the hidden and in the output layers. In this research three structures were tried:

PP = a linear Transfer Function (TF) is used in both, the hidden and the output layers.

LP = a sigmoidal TF in the hidden layer and a linear TF in the output layer

TT = a hyperbola tangent as the TF in the hidden layer and a linear TF in the output layer

The selected transfer functions are defined as follows:

linear function:

$$f_1(n) = n = wp + b \qquad \textbf{(9)}$$

sigmoidal function:

$$f_2(n) = \frac{1}{1+e^{-n}} \qquad \textbf{(10)}$$

hyperbolic tangent function:

$$f_3(n) = \frac{e^n - e^{-n}}{e^n + e^{-n}} \qquad \textbf{(11)}$$

where *n* represents the net input into the network, *w* is the weight matrix, and *b* is the bias vector. Thus, the structures of the neural network change depending on the natural relationship of the available data.

The structure of the network can be represented by a diagram introduced by Hagan (1996). Figure 1 shows that the identified neural network structure for one of the selected cases has a single neuron in both the hidden and the output layers with a linear transfer function in both layers. In this case *L* interaction columns were input into the network and a single interaction constants is the element of the target vector, where *L* is the total number of compounds that content at least one of the interaction of interest.

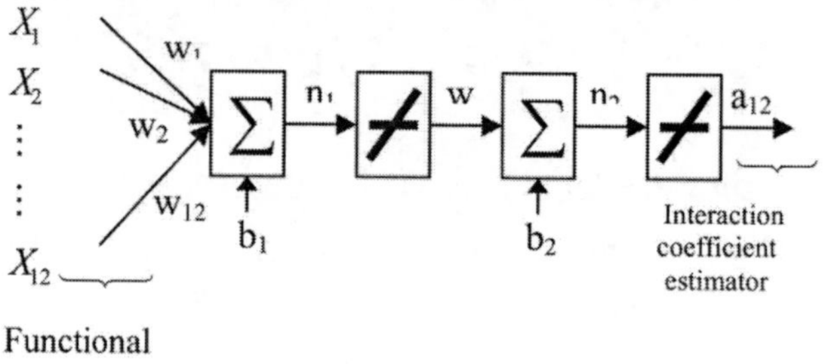

Figure 1. Neural Network Structure

In Figure 1 the *w's* are the weight matrices, *b's* are the biases, *n's* are the net input and a_{12} is the estimated of the interaction constant, A_{12}. It was identified that one of the best learning procedures for the underlying problem was obtained with the Levenberg-Marquradt algorithm (Ramirez-Beltran and Jackson, 1999; Hagan, et. al., 1996).

A parallel effort was conducted to estimate the constant α_{ij} of the NRTL model. The estimation procedure was conducted after using regression and neural network methodology. Since the NRTL constant is a function of temperature and interaction constant, the postulated regression model is as follows:

$$\alpha_{ij} = a_{ij}A_{ij} + a_{ji}A_{ji} + b_1\tau + b_2\tau^2 \tag{12}$$

where A_{ij} are the interaction constants that were previously defined and τ is the temperature at which the equilibrium experiment was conducted, *a's* and *b's* are the regression coefficients to be estimated from library of phase-equilibrium data. A neural network was designed to estimate the NRTL constant, α_{ij}. The training patterns used to identify the second neural network model are the following:

$$P' = [A_{ij} \quad A_{ji} \quad \tau] \qquad \textbf{and} \qquad T = [\alpha_{i,j}] \tag{13}$$

The minimum variance criterion was used to identify the appropriate neural network model. It was found that the neural network that best represents the input output relationship has a two-layer neural network with a single linear transfer function in each layer. The Levenberg-Marquartd algorithm was also used to accomplish the training task. Once the NRTL constant are known the activity coefficients are computed by using equation (1). In summary, the proposed prediction process consists of four major tasks:

Task 1. Generate the information matrix that contains the interactions from all functional groups and the columns associated with interaction constants of interest.

Task 2. Design a neural network to estimate the interaction constants, *A's*.

Task 3. Design a second neural network to estimate the NRTL constant, α.

Task 4. Use the NRTL model, equation (1), to estimate the activity coefficients.

A computer program was developed to perform the proposed estimation method for the activity coefficient. An additional subroutine was also developed to perform identification and estimation of the nonlinear regression model, with the purpose of comparing the performance of the neural network with the regression approach. The computer program is available from the author upon request.

NUMERICAL EXAMPLE

For illustration purposes, consider the estimation of the activity coefficients at 60 °C of a solution formed by ethanol and 1-2-dichloroethane. The ethanol mole fraction varies from 0% to 100% (the complement to 100% is the 1-2-dichoroethane mole fraction). The estimation procedure consists of seven steps:

Step 1. Structural formula. Write the structural formula for each compound of the mixture. Figure 2 shows the structural formula for the considered mixture.

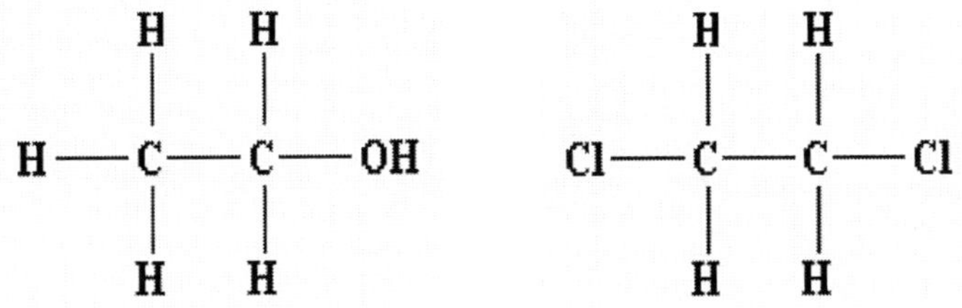

Ethanol **1-2-Dichloroethane**

Figure 2. Structural formula

Step 2. Identification of Functional groups. Breakdown each compound into the corresponding functional groups.

- Ethanol has one unit of the functional group 1, (CH_3); one unit of the functional group 2, (CH_2); and one unit of the functional group 32, (OH). This information is represented by one G_1, one G_2, and one G_{32}. Thus, $a_1 = 1$, $a_2 = 1$, and $a_{32} = 1$, $w_1 = 15.03$, $w_2 = 14.03$, and $w_3 = 17.01$.
- 1-2-Diclhloroethane has the following functional groups: two units of the functional group 72, (CH_2Cl), which is represented by two G_{72}. That is, $a_{72} = 2$, and $w_{72} = 49.48$.
-

Step 3. Develop Information matrix. Develop the information matrix associated to the considered interactions. Each column of the information matrix represents a functional group interaction. Each element of the information matrix includes the number of functional-group units and the molecular weights. Regression and neural network models use the information matrix. The nonlinear regression model exhibits the required columns in the information matrix. For the considered case the regression model is as follows:

$$A_{1,2} = \beta_{1,72} X_{1,72}^{m_{1,72}} + \beta_{2,72} X_{2,72}^{m_{2,72}} + \beta_{32,72} X_{32,72}^{m_{32,72}} + R \quad (14)$$

$$A_{2,1} = \beta_{72,1} X_{72,1}^{m_{172,1}} + \beta_{72,2} X_{72,2}^{m_{72,2}} + \beta_{72,32} X_{72,32}^{m_{72,32}} + R \quad (15)$$

where $A_{1,2}$ represents the NRTL constant between the ethanol and 1-2-dichloroethane, $A_{2,1}$ represents the NRTL constant between the 1-2-dichloroethane and ethanol, R represents other interactions that are included in the model, but they do not belong to the product under interest. For instance, the variable $X_{1,2}$ represents the interaction between the functional group 1 and 2. This information is used to train the network; however, it is not used to perform neural network prediction. On the other hand, the interaction of interest are (1,72), (2,72), (32,72), (72,1), (72,2) and (72,32). The variable $X_{1,72}$ is the functional-group coefficient and represents the interaction between the functional group 1 and 72, and equation (4) was used to estimate its corresponding coefficient. That is, $X_{1,72}$ can be computed as

$$X_{1,72} = a_1 a_{72} w_1 / w_{72} = 1(1)15.03/49.48 = 0.3038.$$

Equation (3) is used to estimate the column where the value $X_{1,72}$ will be located into the information matrix. It should be noted that the number of considered functional groups is 114. Thus, the corresponding column is

$$n_{2,1} = 114(1-1) + 72 = 72.$$

Similarly,

$$X_{2,72} = 0.8506, \qquad n_{2,72} = 186,$$
$$X_{32,72} = .6876, \qquad n_{32,72} = 3606,$$
$$X_{72,1} = 6.5842, \qquad n_{72,1} = 8095,$$
$$X_{72,2} = 7.0535, \qquad n_{72,2} = 8096, \text{ and}$$
$$X_{72,32} = 3.4077, \qquad n_{72,32} = 8{,}126.$$

Thus, the layout of the information matrix associated with A_{12} is exhibited in Table 1.

Table 1. The matrix information for A_{12}

A_{ij}	1	2	72	115	116	186	3606	10033	10134	10104
-335.2038	2.1426	0.3038	0.3038	2.8004	6	0.8506	0	5.3160	11.3899	1.6148
17.6153	0	0	0.6075	0	0	0.5671	0.6876	0	0	0
183.2452	0	0	0.6075	0	0	0.5671	0.6876	0	0	0
215.4029	0	0	0.6075	0	0	0.5671	0.6876	0	0	0
⋮	⋮	⋮	⋮	⋮	⋮	⋮	⋮	⋮	⋮	⋮
133.2760	0	0	0.6075	0	0	0.5671	0.6876	0	0	0
162.3080	0	0	0.6075	0	0	1.1342	0.6876	0	0	0

The available phase-equilibrium data provide information for the following columns: 1, 2, 72, 115, 116, 186, 3606, 8095, 8096, 8126, 10033, 10034, and 10104. The columns of interest for the interaction constant A_{12} are 72, 186, 3606, and the columns of interest for the constant A_{21} are 8095, 8096, and 8126. It should be noted that the information matrix includes other interactions that do not correspond to the columns of interests; for instance, the columns 1, 2, 115, etc. It should be noted that the first row for the column 3606 has a zero value, indicating that the selected compound in the equilibrium data does not have that particular functional group interaction. The interactions exhibited in Table 1 are used to train the neural network because the A's constants are associated with the corresponding interactions. However, when the A's constants will be predicted the network is evaluated only with the interactions of interests. It should be noted that Table 1 does not contain the columns 8095, 8096, and 8126, since this table was designed for the constant A_{12}. Thus, when the interaction for A_{21} is extracted then the columns 8095, 8096, and 8126 will be included in the information matrix.

Step 4. Generate the training patterns. The training patterns are organized to perform the neural network-training task. The input patterns are represented with P_t and the targets are expressed by T_t, where t represents the number of available training patterns. That is,

$$P_1 = \left[x_{1,1}\; x_{1,2} \quad x_{1,72} \quad x_{1,115} \quad \cdots \quad x_{1,10104}\right] \quad \text{and} \quad T_1 = \left[A_{1,1,2}\right]$$
$$P_2 = \left[x_{2,1}\; x_{2,2} \quad x_{2,72} \quad x_{2,115} \quad \cdots \quad x_{2,10104}\right]' \quad \text{and} \quad T_2 = \left[A_{2,1,2}\right]$$
$$\vdots$$
$$P_t = \left[x_{t,1}\; x_{t,2} \quad x_{t,72} \quad x_{t,115} \quad \cdots \quad x_{t,10104}\right] \quad \text{and} \quad T_t = \left[A_{t,1,2}\right]$$

Step 5. Estimation of the interaction constants.
The proposed *minimum-variance estimation criterion* is implemented to identify the appropriate neural network model for the underlying system. The identification procedure can be described as follows.

1) The structure PP was selected and the training task was conducted with the purpose of recording the sse and the estimated $\hat{A}_{12}$ values. This process was repeated five times with different initial weight and bias matrices. Since a sample size of five has been identified to provide enough information to show whether or not the neural network estimation is stable. The typical representation of the learning process is a plot between the number of epochs versus the sse. Figure 3 exhibits that the computer routine indicates that the gradient becomes zero after 9 epochs indicating that weights associated with a local minimum were found after 9 iterations. At the end of five replications the sample variance of $\hat{A}_{12}$ was computed.
2) The number of neurons in the hidden layer was increased by one unit and again five training replicates were conducted. The sse and the $\hat{A}_{12}$ values were recorded, as well as the sample variance of $\hat{A}_{12}$ was computed. This process was repeated until the minimum of the sample variance of $\hat{A}_{12}$ was identified. In the considered case five neurons in the hidden layer were sufficient to identify the minimum variance estimator. Table 2 shows the averages of the sse and the sample variance of the estimates.
3) A different neural network structure is selected, for instance LP, and the process is repeated. Table 2 also summarizes the calculus for three neural network structures.
4) The best structure is the one that provides *the minimum variance with the least number of neurons*. Table 2 indicates that the best structure for the underlying problem is the PP, i.e., the best structure for the given case includes a single neuron with linear transfer function in both the hidden and in the output layers.

Similarly, Table 3 exhibits computational results to identify the best neural network structure to estimate A_{21}. This table shows that the best neural network structure for estimating A_{21} is also PP, with a single neuron in both the hidden and the output layers.

Table 2. Identification Results for System 1 and Parameter A_{12}

Structure	Neuron in the hidden layer	Average(sse)	Min{Var(A)}	Estimate, A_{12}
PP	1	125,410	0	117.67
PP	2	125,410	0	117.67
PP	3	125,410	0	117.67
PP	4	125,410	0	117.67
PP	5	125,410	0	117.67
LP	1	181,150	766.89	unstable
LP	2	181,150	3,166.70	unstable
LP	3	125,410	0	117.67
LP	4	125,410	0	117.67
LP	5	181,150	3,166.70	unstable
TP	1	181,150	3,166.70	unstable
TP	2	231,490	3,936.60	unstable
TP	3	181,150	3,166.70	unstable
TP	4	125,410	0	117.67
TP	5	125,410	0	117.67

Table 3. Identification Results for System 1 and Parameter A_{21}

Structure	Neuron in the hidden layer	Average(sse)	Min{Var(A)}	Estimate, A_{12}
PP	1	357,100	0	1,254.9
PP	2	357,100	0	1,254.9
PP	3	357,100	0	1,254.9
PP	4	357,100	0	1,254.9
PP	5	357,100	0	1,254.9
LP	1	643,100	160,000	unstable
LP	2	1,092,300	240,000	unstable
LP	3	821,400	170,000	unstable
LP	4	971,200	110,000	unstable
LP	5	968,100	240,000	unstable
TP	1	1,120,900	0	1,145.9
TP	2	971,200	110,000	unstable
TP	3	1,120,900	0	1,145.9
TP	4	789,800	310,000	unstable
TP	5	821,400	170,00	unstable

Five systems were arbitrarily selected from DECHEMA data (Sorense and Arlt, 1979) to perform comparison between the neural network approach and regression method. The interaction constants, A_{12} and A_{21} were estimated using the neural network identification procedure. Results from regression and neural network models were compared and it was found that the neural network outperforms the regression method. Results are presented in Table 4, which summarizes the NRTL prediction constants for the five systems. The first system is referred to the ethanol and 1-2-dichloroetane mixture. The system 2 consists of methanol with triethylamine, system 3 refers to methyl acetate with methanol, system 4 consists of methanol with ethyl acetate, and system 5 is composed of 1-propanol with chlorobenzene.

Step 6. Estimation of the alpha constant. Obtain an estimator for alpha parameter based on the interaction constants, *A*'s. The alpha parameter, α_{ij}, is estimated by using a polynomial function of *A's* and temperature, assuming that at this stage the values of *A's* are known. Regression model, equation (12), was used to estimate the alpha constant and results are exhibited in Table 4. It should be noted that a neural network was also designed to estimate the values of alpha and predictions are also shown in Table 4.

Step 7. Estimation of the activity coefficient. Once the values of *A's* and α are predicted, equation (1) is used to predict the activity coefficients. It should be noted that the mole fraction for each compound and the temperature of the system are also needed to estimate the activity coefficients. The estimates of the activity coefficients for the ethanol and 1-2-diclhoroethane are exhibited in Table 5. The variables x_1 and x_2 represents the mole fraction of the ethanol and 1-2-dichloroetane, respectively. The next column of Table 5 refers to the activity coefficients that were estimated based on equilibrium experimental data. The last two columns exhibit the activity coefficients predicted based on regression, and neural network methodology, respectively. It should be noted that, regression and neural network approaches do not use the phase-equilibrium data, they use the functional group concept to predict the NRTL constants and consequently activity coefficients are predicted. The accuracy of the neural network prediction is exhibited in Figure 4. The continuous lines represent the activity coefficients based on the equilibrium experimental data and the broken lines represent the prediction of activity coefficients based on the neural network methodology. The first activity coefficient is associated with ethanol and varies in the range from 1 to 60, and the second activity coefficient is associated with 1-2-diclhoroethane, which varies from 1 to 5. The actual values of the activity coefficients are presented in Table 5. It can be noted that the mole fraction of the ethanol varies from 0 to 1. Figures 5-8 exhibit the neural network prediction of the activity coefficients for the systems 2, 3, 4, and 5, respectively. The considered temperatures are 40° C, 40° C, 40° C, and 95° C for the systems 2, 3, 4, and 5, respectively. The quality of the prediction is measured by the average mean absolute deviation. That is,

$$e_j = \frac{1}{n}\sum_{i=1}^{n}\left|\gamma_{i,j} - \hat{\gamma}_{i,j}\right| \qquad \textbf{(16)}$$

where n is the number of mole fraction levels used, e_j is the average mean absolute deviation for the jth compound, $\gamma_{i,j}$ and $\hat{\gamma}_{i,j}$ are the activity coefficients obtained from experimental data and from the neural network methodology, respectively. The subscript i refers to the applied mole fraction level and the subscript j represents the chemical compound. The computed averages for each of the systems are exhibited in Table 6. For instance, in the first system $e_1 = 1.1147$ and $e_2 = 0.3180$ are the mean absolute deviation for the first and the second activity coefficient, respectively. Thus, there is a smaller error in estimating the activity coefficient for the ethanol than for the 1-2-dichloroetane.

CONCLUSIONS

A neural network methodology is proposed to predict the NRTL interaction constants based on the knowledge of the structural formula for the components of a given mixture. This approach exhibits a potential tool to predict the activity coefficients of chemical compounds for which experimental data are not available.

The minimum variance criterion is proposed to identify the appropriate structure of a feedforward neural network. The introduced methodology basically consists of selecting a neural network structure with a single neuron in the hidden layer, and then increase the number of neurons until the minimum variance is identified. Finally, the neural network structure that provides the minimum variance with the least number of neurons is selected as the best neural network structure to properly represents the underlying system.

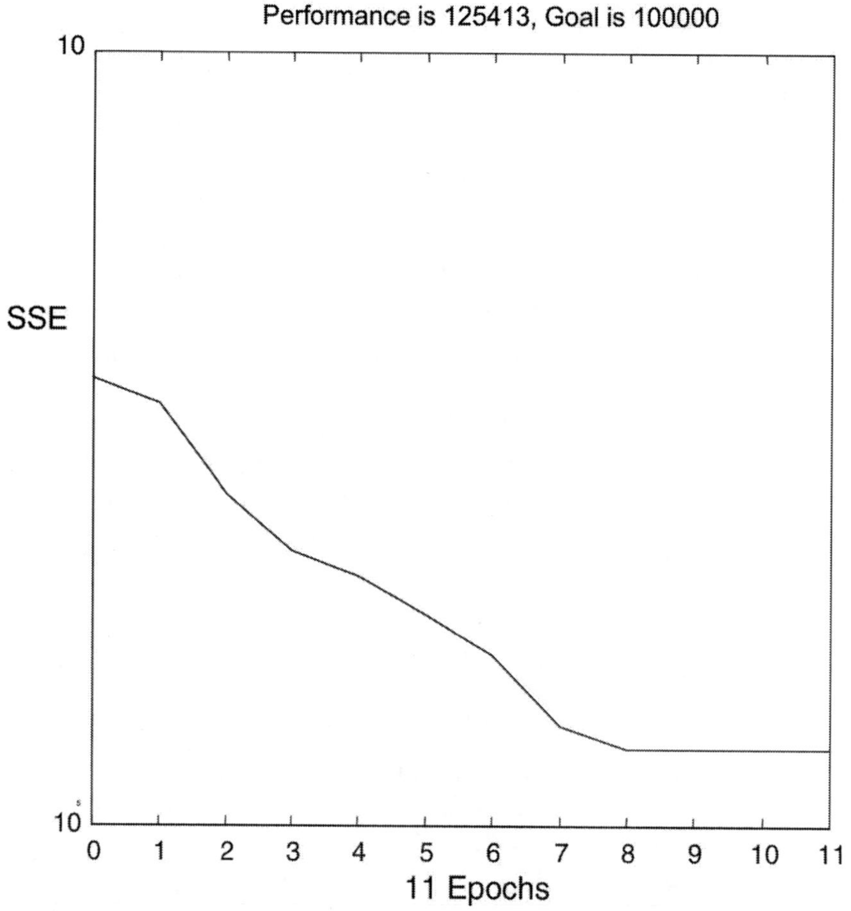

Figure 3. The Training Process

Table 4. Interaction Constant Estimation

Systems	Constants	Estimation from Equilibrium Experimental Data	Estimation from Regression approach (using functional groups)	Estimation from Neural Networks approach (using functional groups)
	Case 1			
Ethanol	A_{12}	215.4029	32.6742	117.6757
1-2-dechloroethane	A_{21}	1051.0773	914.427	1254.9
	α	.2898	.2913	.2923
	Case 2			
Methanol	A_{12}	-443.8352	25.8128	-352.20
Triethylamine	A_{21}	973.8281	774.902	894.21
	α	.2883	.3256	.3018
	Case 3			
Methyl Acetate	A_{12}	261.4306	106.387	415.39
Methanol	A_{21}	470.7231	140.016	380.99
	α	.2964	.3387	.3063
	Case 4			
Methanol	A_{12}	609.6904	426.582	452.2663
Ethyl Acetate	A_{21}	250.4283	90.3626	167.53
	α	.2960	.3363	.3098
	Case 5			
1-propanol	A_{12}	382.4952	685.3570	382.49
chlorobenzene	A_{21}	656.2518	318.2130	656.25
	α	.2951	.21763	.3015

Table 5. Estimation of Activity Coefficient

Mole fraction		Activity Coefficient Estimated from Experimental Data		Activity Coefficient Estimated from Regression Model (using functional groups)		Activity Coefficient Estimated from Neural Networks (using functional groups)	
x_1	x_2	γ_1	γ_2	γ_1	γ_2	γ_1	γ_2
0	1.0000	40.0871	1.0000	17.1176	1.0000	59.4624	1.0000
0.0417	0.9583	21.4795	1.0130	10.6944	1.0098	25.5594	1.0174
0.0833	0.9167	12.9327	1.0475	7.2387	1.0362	13.3982	1.0616
0.1250	0.8750	8.5274	1.0992	5.2278	1.0760	8.1096	1.1249
0.1667	0.8333	6.0405	1.1656	3.9806	1.1270	5.4596	1.2031
0.2083	0.7917	4.5308	1.2453	3.1657	1.1880	3.9824	1.2936
0.2500	0.7500	3.5588	1.3378	2.6099	1.2581	3.0887	1.3948
0.2917	0.7083	2.9023	1.4428	2.2172	1.3365	2.5125	1.5057
0.3333	0.6667	2.4412	1.5607	1.9315	1.4228	2.1218	1.6257
0.3750	0.6250	2.1066	1.6919	1.7186	1.5167	1.8461	1.7543
0.4167	0.5833	1.8572	1.8373	1.5568	1.6181	1.6453	1.8916
0.4583	0.5417	1.6671	1.9981	1.4320	1.7266	1.4952	2.0374
0.5000	0.5000	1.5197	2.1755	1.3344	1.8423	1.3808	2.1921
0.5417	0.4583	1.4037	2.3714	1.2574	1.9651	1.2921	2.3558
0.5833	0.4167	1.3114	2.5879	1.1963	2.0949	1.2226	2.5292
0.6250	0.3750	1.2375	2.8275	1.1477	2.2318	1.1677	2.7126
0.6667	0.3333	1.1780	3.0930	1.1090	2.3756	1.1242	2.9069
0.7083	0.2917	1.1301	3.3881	1.0783	2.5266	1.0897	3.1128
0.7500	0.2500	1.0919	3.7171	1.0542	2.6846	1.0626	3.3313
0.7917	0.2083	1.0617	4.0848	1.0357	2.8499	1.0415	3.5636
0.8333	0.1667	1.0383	4.4972	1.0217	3.0225	1.0255	3.8108
0.8750	0.1250	1.0211	4.9615	1.0116	3.2024	1.0138	4.0745
0.9167	0.0833	1.0092	5.4863	1.0049	3.3899	1.0059	4.3563
0.9583	0.0417	1.0023	6.0817	1.0012	3.5850	1.0014	4.6580
1.0000	0	1.0000	6.7602	1.0000	3.7879	1.0000	4.9818

Table 6. Mean Absolute Deviation for the Activity Coefficients

Case	Neural Network Results		Regression Results	
	First Deviation (e_1)	Second Deviation (e_2)	First Deviation (e_1)	Second Deviation (e_2)
1	1.1147	0.3180	1.9997	0.7039
2	0.1509	0.1033	1.3888	0.8215
3	0.1301	0.3312	1.1344	0.9472
4	0.5335	0.8399	0.7862	1.1319
5	0.0174	0.0187	0.5927	0.4927

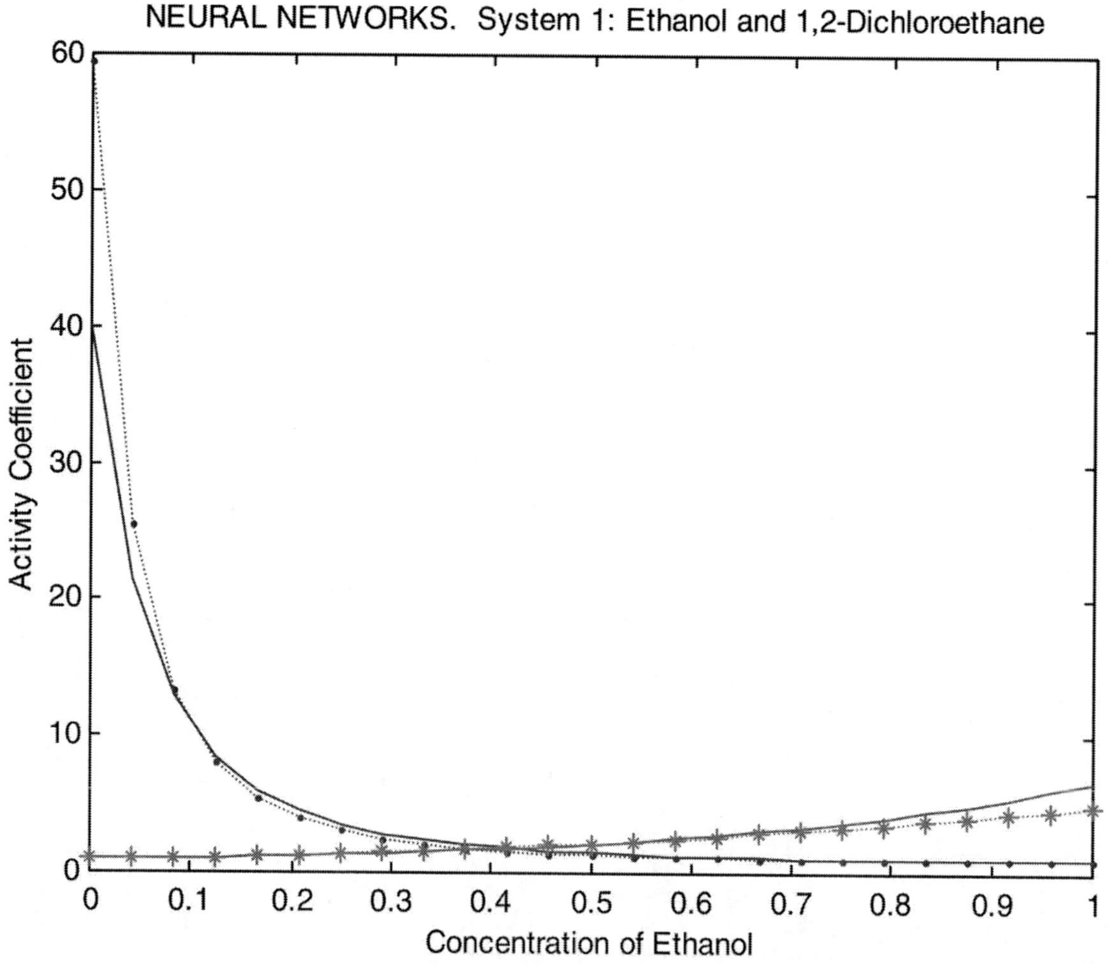

Figure 4. System 1: Ethanol and 1,2-Dichloroethane

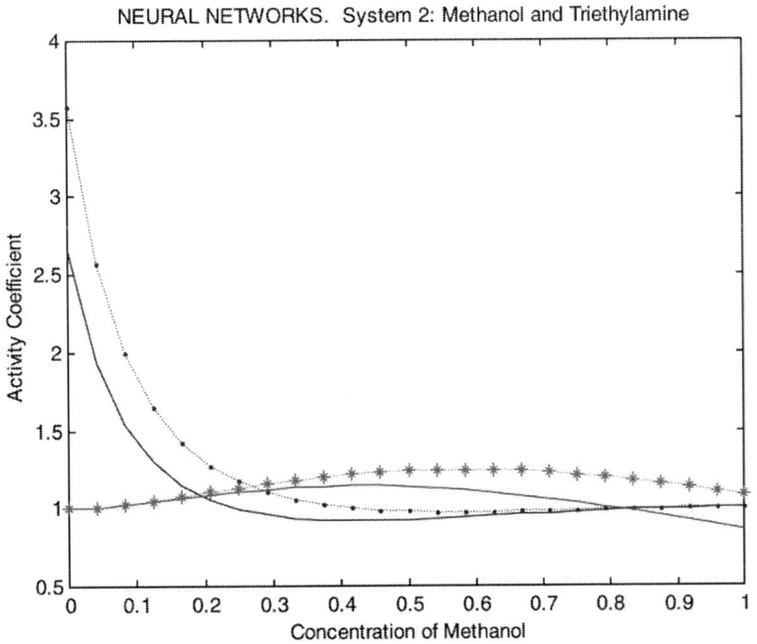

Figure 5. System 2: Methanol and Triethylamine

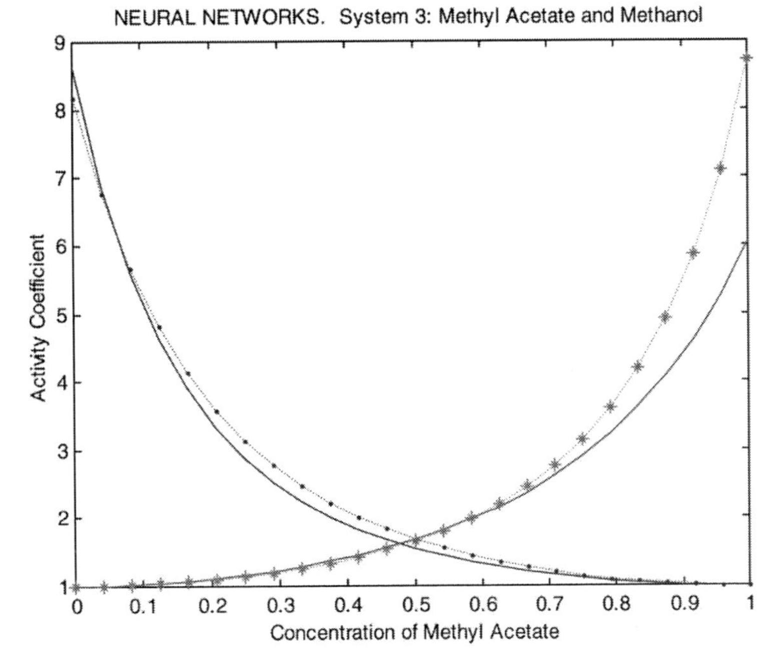

Figure 6. System 3: Methyl Acetate and Methanol

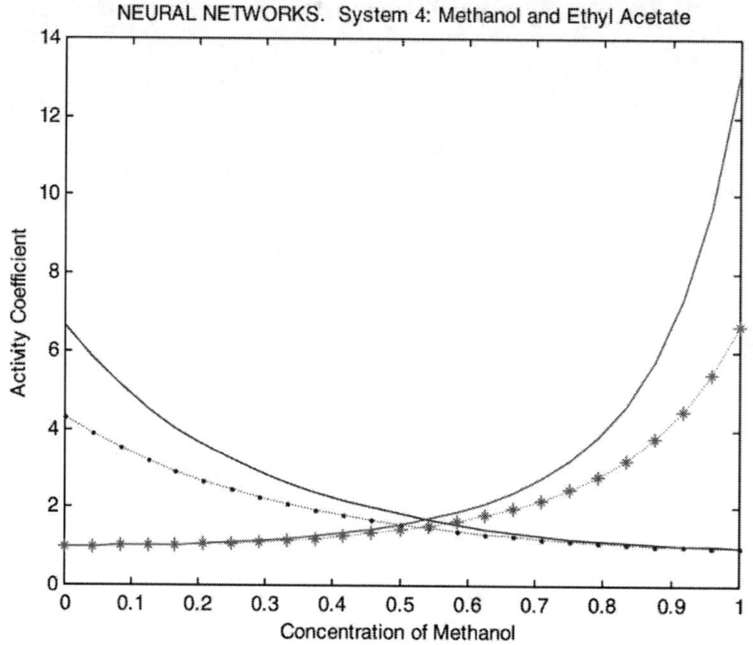

Figure 7. System 4: Methanol and Ethyl Acetate

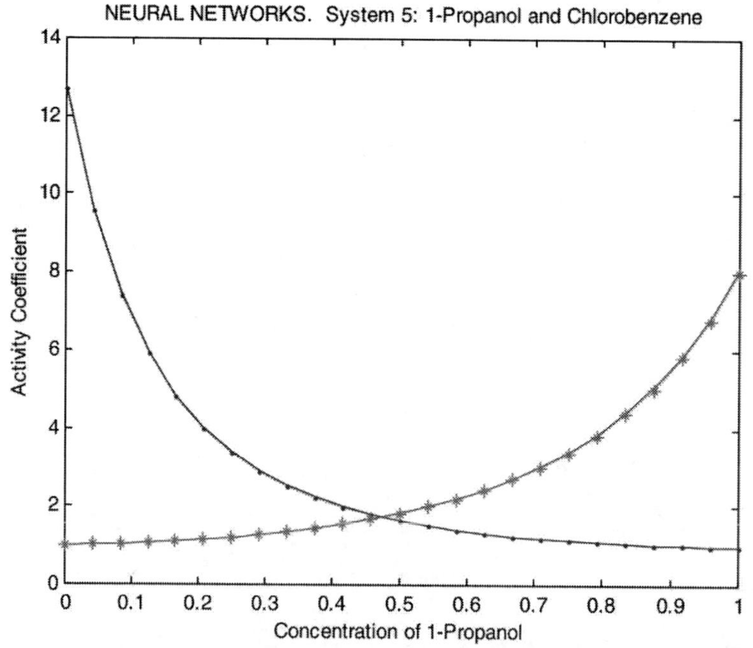

Figure 8. System 5: 1-Propanol and Chlorobenzene

REFERENCES

Abrams, D.S., and Prausnitz, J.M., Statistical Thermodynamics of Liquid Mixtures: A New Expression for the Excess Gibbs Energy of Partial or Completely Miscible Systems. *AIChE Journal*, Vol. 21 No. 1 116-128, (1975).

Bazaraa, M.S., Sherali, H.D., and Shetty, C.M., *Nonlinear Programming: Theoty and Algorithms*, 2nd Ed., John Wiley & Sons, Inc., New York, (1993).

Box, G.E.P., and Jenkins, G.M., *Time Series Analysis: Forecasting and Control*, Holden-Day, Oakland CA, (1976).

Cartensen,J.T., *Drug Stability, Principles and Practice.* Second Ed., Marcel Dekker Inc., (1995).

Constantinou, L., and Gani, R., New Group Contribution Method for Estimating Properties of Pure Compounds, *AIChE Journal*, Vol. 40, No. 10, 1697-1710, (1994).

Fredenslund, A., Jones, R.L., and Prausnitz, J.M., Group Contribution Estimation of Activity Coefficients in Nonideal Liquid Mixtures, *AIChE Journal*, Vol. 21, No. 6, 1086-1099, (1975).

Hagan, M.T., Demuth, H.B., and Beal, M., Neural Network Design, *PWS Publishing Company*, Boston, (1996).

Pandit, S.M. and Wu, S.M., *Times Series and System Analysis with Applications,* John Wiley and Sons, New York, (1983).

Ramirez-Beltran, N.D., Sridhar, L.N., and Rodriguez, H., A Nonlinear Model to Estimate Chemical Interaction Constants, *Computing Research Conference,* CRC'98, December 4, Mayaguez, Puerto Rico (1998).

Ramirez-Beltran, N.D., and Jackson, H., Application of Neural Networks to Chemical Process Control, *Computers and Industrial Engineering.* Vol. 37, Nos. 1-2, 387-390, October (1999)

Smith, J.M., Van Ness, H.C., and Abbott, M.M., *Introduction to Chemical Engineering Thermodynamics*, McGraw-Hill, New York, (1996).

Sorensen, J.M., and Arlt, W., DECHEMA Chemistry Data Series, Frankfurt, (1979).

Wei, W.W.S., *Time Series Analysis: Univariate and Multivariate Methods,* Addison-Wesley Publishing Co., Redwood City, CA, (1990).

Controlled Particle Deposition by Design of an Electrochemical Adsorption Cell

Ehsan Bakhshi, R&D Center of National Petrochemical Company, Tehran, Iran, Fatemeh Abniki, National Petrochemical Company,Ghadir Group, Tehran, Iran, and Majid Mosalla, Metallurgy & Material Science Department of Shiraz University, Tehran, Iran.

Deposition of colloidal particles substrate is encountered in a variety of naturally occurring processes such as particulate fouling of heat exchangers, thrombus formation in vascular prostheses and other artificial organs and aerosols entering the lungs. Deep bed filtration of waste waters is perhaps the most prominent example of importance to industry. Design of an electrochemical adsorption cell utilizing reticulated vitreous carbon (R.V.C.) as the working electrode ,stainless steel as the counter electrode and a cellulose acetate membrane ,for separating the anodic and cathodic, is described for the separation of 5.4 m polystyrene latex colloidal particles from a KCI solution. The experimental results are compared with the Derjaguin-Landau-Verwey-Ovdrbeek (D.L.V.O.) prediction for the occurrence of favorable and unfavorable deposition conditions. In the theoretical section of this work it will be assumed that the main energies contributing adsorption interaction between the surfaces will be due to electrical double layer to Van der Waals attraction. The some of these energies provides the total interaction energy between surfaces and this constitutes the underlying assumption for the classical D.L.V.O. theory of colloidal stability. The polystyrene latex particles used in this study were 5.4 μm and hence effect of diffusion becomes negligible. For reasonably low flow rates, Re

Web Page: www.nipc.net

The Effects of Confinement on Water Structure and Dynamics: a Molecular Simulation Study

Pussana Hirunsit and Perla B. Balbuena. Chemical Engineering, Texas A&M University, 3122, TAMU, College Station, TX 77843-3122

Water confined in carbon nanostructures interestingly shows an ordered, ice-like structure. These structures can be potentially used as proton-conductor media. Such ice-like structure may have however different structural and dynamical properties than those of ice. We performed classical molecular dynamics simulations to characterize the structure and dynamics of water at room temperature and a density of 1 g/cm^3 within graphite walls of different separations, and establish differences between structural and dynamical properties of confined water and those of low-temperature water at the same density that could be useful to describe the environment in which a proton would move under these conditions. The results show the formation of well-ordered and parallel to the graphite surfaces of water layers. For larger separations additional water layers are formed in the pore center, and the diffusion coefficient increases. The mobilities of encapsulated water become similar to the one of low-temperature water (213 K) at a separation of 8 Å although they show very different structures.

An Invesitgation of Protein Adsorption on Zinc Oxide

Liqiang Zheng, Lei L. Kerr, and Marvin E. Thrash Jr. Paper and Chemical Engineering, Miami University, 240 Gaskill Hall, Oxford, OH 45056

The adsorptive behavior of BSA and Urease onto a zinc oxide adsorbent was investigated. The dependence of protein surface concentration on pH was studied. Isotherm measurements indicated that protein adsorption significantly varied with pH. Protein adsorption was essentially non-existent at elevated pH, however, the surface concentration of protein steadily increased when the acidity level of the buffer was raised. At a pH value of 5, multilayer protein adsorption was observed for BSA. Zeta potential measurements indicated that the protein and the zinc adsorbent each possessed a strong negative potential at high pH, however, the magnitude of negative potential was reduced at lower pH. As the solution became more acidic, the driving forces associated with hydrogen bonding and hydrophobic interactions appeared to overcome the repulsive forces arising from the negative potential on the protein and the zinc oxide adsorbent. Urease was also adsorbed onto the surface of zinc oxide. The impact of immobilization on Urease will be tested by studying the enzymes ability to degrade urea. The urea degradation kinetics of adsorbed Urease will be compared to the degradtion kinetics of unbound Urease in solution.

Web Thermo Tables – an on-Line Database of Evaluated Thermochemical Properties for Pure Compounds

Andrei Kazakov, Chris Muzny, Robert D. Chirico, Vladimir V. Diky, and Michael Frenkel. Physical and Chemical Properties Division, NIST, 325 Broadway, Boulder, CO 80305

It has long been understood that experimental thermophysical and thermochemical property data are vital to scientific research and industrial design. For over 50 years, Thermodynamics Research Center (TRC) has been publishing the tables of critically-evaluated data covering physical and thermodynamic properties of pure compounds, TRC Tables-Hydrocarbons and TRC Tables-Non-Hydrocarbons. Over their long history, the TRC Tables have always been valued as reputable source of evaluated thermophysical and thermodynamic data. Starting from 2004, TRC was also issuing annual releases of the electronic version of these tables, WinTable, as a standalone Windows software package. To facilitate more flexible, convenient, and up-to-date access to the data, here, we present the release of on-line version of the TRC tables, Web Thermo Tables (WTT). WTT will be distributed to the users using Web-Oracle dissemination system recently established at NIST/Boulder. Presently, WTT contains data for 7729 compounds, and over 800,000 individual data points. The tabulated information includes critical properties, vapor pressures and boiling temperatures, phase transition properties, volumetric properties, heat capacities and derived properties, transport properties, reaction state-change properties, as well as index of refraction, surface tension, and speed of sound. Various search options and data plotting capabilities are provided via Web interface. WTT will be distributed through the NIST Standard Reference Database Program.

Evaluating and Fine Tuning Cubic Equations of State For Supercritical Water Processes

Shyh-Ming Chern, I-Shou University, Kaohsiung, Taiwan, ROC

ABSTRACT

Numerous equations of state have been developed over the years and new ones are continuously being proposed. The performance of these models in terms of the accuracy of predicting the state of matters has improved over time. Nevertheless, no perfect model has been found that is capable of predicting the state of matter accurately under all circumstances. The application of equations of state always involves a compromise between the efficiency and the accuracy. More often than not, cubic equations of state, especially with three parameters, are chosen due to their simple mathematical form, being easy to apply, and reasonably good accuracy. In the present study, the performance of the well-known Peng-Robinson and Patel-Teja equations of state has been examined in terms of their ability to predict the behavior of water around the critical area, including the saturated region from 0.01°C to critical point and the superheated region (10-60 atm, 400-600°C) . The measures compared include the specific volume, the enthalpy departure and the entropy departure. Both EoS's work reasonably well in terms of predicting the specific volume with Patel-Teja's EoS being clearly the better of the two. However, in terms of predicting the enthalpy departure and entropy departure the two EoS's are much more matched in most cases. The highest discrepancy observed always occurs at the critical point, which is about 47% for the specific volume, 28% for the enthalpy departure and 43% for the entropy departure. Using an acentric factor of 0.5491 instead of its genuine value in Peng-Robinson EoS for water, reduces the discrepancies dramatically, generally to about one third or less. However, it is difficult to retune PT-EoS for water.

Keywords: equation of state, supercritical water, EoS, SCW

INTRODUCTION

Numerous organic matters have been reacted in supercritical water, e. g. alkanes[1], organic solvents[2], organic wastes[3], glucose, lignin, biomass, etc. In a recent study, Tang and Kitagawa[4] simulated the gasification of biomass using Peng-Robinson EoS[5] based on the assumption of thermodynamic equilibrium and the method of free energy minimization. Since an EoS capable of predicting the state of water and the associated thermodynamic properties with good enough accuracy is the key to the successful modeling of various processes undergoing in high temperature, high-pressure water, it is desirable to evaluate the performance of an EoS prior to its implementation in a simulation. In the present study, the performance of the well-known, well-developed Peng-Robinson and Patel-Teja[6] equations of state has been examined in terms of their ability to predict the behavior of water around the critical area, including the saturated region from 0.01°C to critical point and the superheated region (10-60 atm, 400-600°C) . The measures compared include the specific volume, the enthalpy departure and the entropy departure. After the examination, the parameters of the EoS are adjusted in an attempt to better predict the water behavior near its supercritical point.

EQUATIONS OF STATE

Numerous equations of state have been developed over the years and new ones are continuously being proposed. The performance of these models in terms of the accuracy of predicting the state of matters has improved over time. Nevertheless, no perfect model has been found that is capable of predicting the state of matter accurately under all circumstances. The application of an EoS always involves a compromise between the efficiency and the accuracy. More often than not, cubic equations of state, especially with three parameters, are chosen due to their simple mathematical form, being easy to apply, and reasonably good accuracy. Since the appearance of van der Waals EoS in 1873,

$$P = \frac{RT}{v-b} - \frac{a}{v^2} \tag{1}$$

many modifications have been proposed, and their performance has improved ever since. Because all these modified van der Waals EoS's can be transformed into third-order polynomial equations, they are termed cubic EoS. The first major modification was presented by Redlich and Kwong[7] by regarding the adjustable parameter of the second term on the right-hand-side of Eq. (1) as a function of temperature

$$P = \frac{RT}{v-b} - \frac{a}{T^{1/2}v^2} \tag{2}$$

Since then, modifications have been focused on the second term of Eq. (1), including those of Peng-Robinson and Patel and Teja EoS's, which will be examined in this study. No comprehensive review on the cubic EoS's is attempted here.

The Peng-Robinson EoS takes the form

$$P = \frac{RT}{v-b} - \frac{a(T)}{v(v+b)+b(v-b)} \tag{3}$$

where

$$a(T) = 0.45724 \frac{R^2 T_C^2}{P_C} \alpha(T, \omega), \tag{4}$$

$$\sqrt{\alpha} = 1 + \kappa(1 - \sqrt{T/T_C}), \tag{5}$$

$$\kappa = 0.37464 + 1.54226\omega - 0.26992\omega^2, \tag{6}$$

which was originally proposed for nonpolar species and has been well developed for a wide variety of pure components and mixtures. The Patel-Teja EoS

$$P = \frac{RT}{v-b} - \frac{a(T)}{v(v+b)+c(v-b)} \tag{7}$$

is very similar to the PR-EoS in form, although the rational behind the development is quite different, with

$$a(T) = \Omega_a \frac{R^2 T_C^2}{P_C} \alpha(T), \tag{8}$$

$$\sqrt{\alpha} = 1 + F(1 - \sqrt{T/T_C}), \tag{9}$$

$$b = \Omega_b R T_C / P_C \ , \tag{10}$$

$$c = \Omega_c R T_C / P_C \ , \tag{11}$$

where, for water, $\Omega_a = 0.50455$, $\Omega_b = 0.065103$, $\Omega_c = 0.1930$ and $F = 0.689803$. The PT-EoS was developed for both nonpolar and polar components, and hence is expected to perform better than the PR-EoS for water. Both PR-EoS and PT-EoS determined their parameters by fitting the equation to the saturation curve from a low pressure to the critical point, resulting in a poor performance at the critical point, as can be seen later. The formulation for the evaluation of thermodynamic properties is well documented for Both PR-EoS and PT-EoS.

EVALUATION OF EQUATIONS OF STATE

To evaluate the performance of EoS's, data from steam stable were employed. For each data point, the experimental enthalpy departure and the entropy departure of the gaseous phase can be derived easily. At the same T and P of the data point, , the compressibility factors were determined for PR-EoS and PT-EoS, and the specific volume, the enthalpy departure and the entropy departure of the gaseous phase were then calculated and compared to the experimental data. The comparisons for the specific volume, the enthalpy departure and the entropy departure along the water saturation curve are presented on Figures 1 through 3, respectively. The counterparts for a typical supercritical area, represented by the region enveloped in 400-600°C and 10-60 MPa, , are presented on Figures 4 through 6. The experimental data from the steam table and their derived values are denoted by solid lines with symbols.

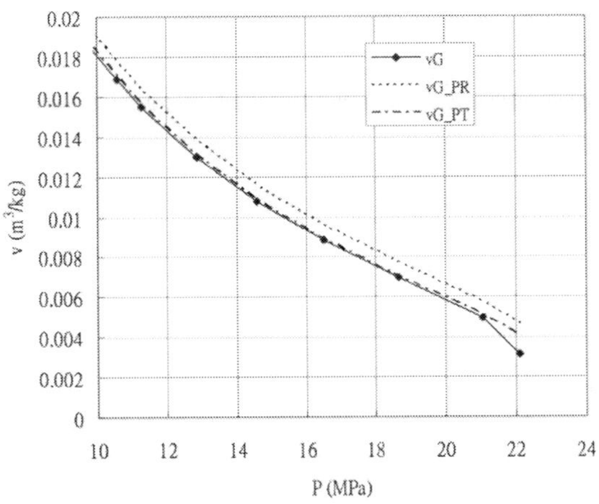

Figure 1. Comparison of specific volumes along the water saturation curve.

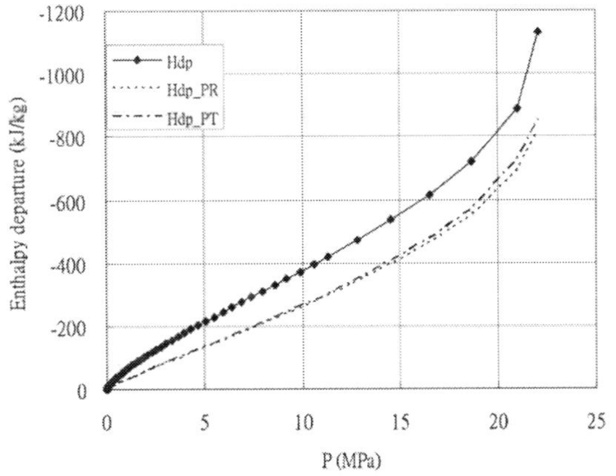

Figure 2. Comparison of enthalpy departures along the water saturation curve.

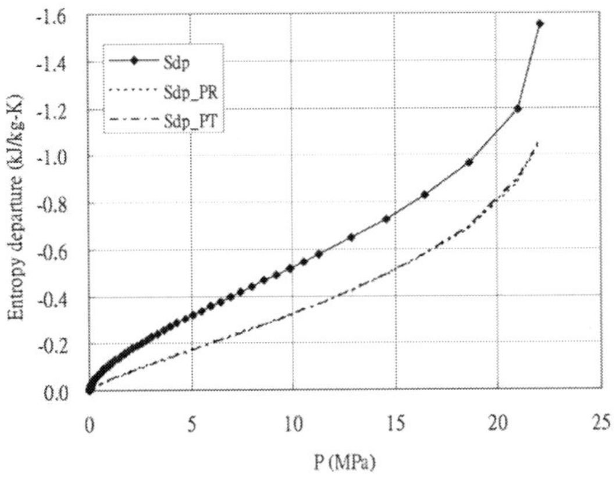

Figure 3. Comparison of entropy departures along the water saturation curve.

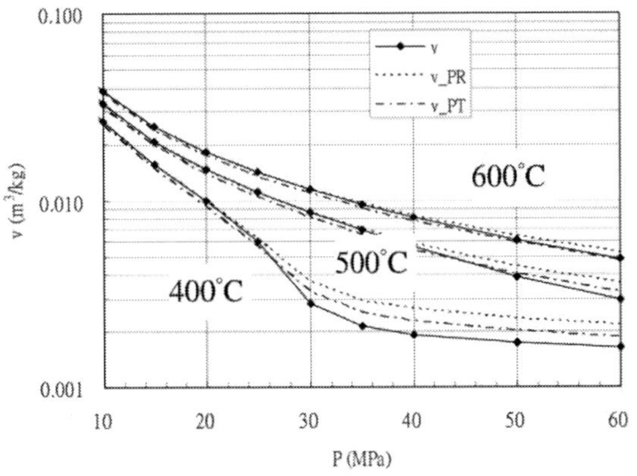

Figure 4. Comparison of specific volumes for the superheated vapor region.

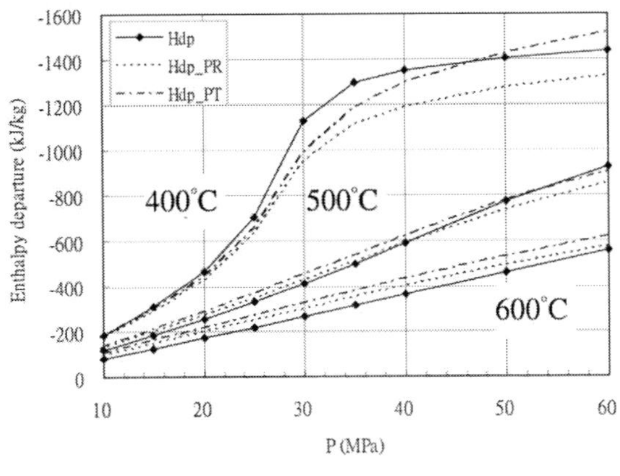

Figure 5. Comparison of enthalpy departures for the superheated vapor region.

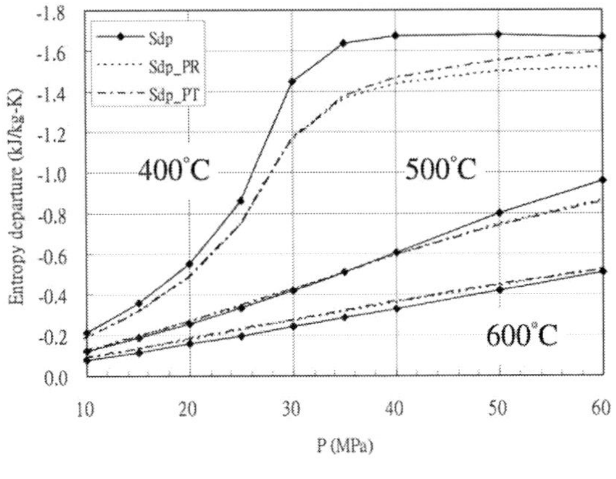

Figure 6. Comparison of entropy departures for the superheated vapor region.

As can be seen from Figure 1, where only the higher pressure portion of the water saturation curve is presented for better clarity, the PT-EoS constantly predicts the specific volume better than the PR-EoS with a minimal discrepancy, except for the vicinity of the critical point. This is understandable since the PR-EoS adopted a fixed Z_C, the compressibility factor at the critical point, of 0.3074 for all substances and the PT-EoS has a Z_C of 0.269 for water, whose genuine Z_C is 0.230. In other words, the performance of the two EoS's at the critical point is sacrificed for the overall performance of the EoS. The performance of the PT-EoS is only slightly better in predicting the enthalpy departures and no better at all in predicting the entropy departures than the PR-EoS along the saturation curve, as seen from Figures 2 and 3. Sizable amount of discrepancies exist even in the low pressure area with an average of about -150 kJ/kg for enthalpy departure and about -0.2 kJ/kg-K for entropy departure. For the supercritical region, the discrepancies for all the three measures begin to shoot up as the pressure increases above 25 MPa and the temperature is relatively low, about 400°C, as seen from Figures 4 through 6. At higher temperature, the discrepancies seem relatively small. The performance of PT-EoS seems to be better in the high pressure and low temperature region but not clearly superior when the temperature is above 500°C, and is even inferior in predicting the enthalpy departure when the temperature is near 600°C.

Both PR-EoS and PT-EoS work reasonably well in predicting the specific volume with Patel-Teja's EoS being clearly the better of the two. However, in predicting the enthalpy and entropy departures the two EoS's are much more matched in most cases. The highest discrepancy observed always occurs at the critical point, about 47% for the specific volume, 28% for the enthalpy departure and 43% for the entropy departure. Given the magnitude of the discrepancy observed, one can only expect predictions with moderate accuracy from the application of the two EoS's to the simulation of supercritical water processes. Therefore, a more dedicated EoS or an EoS better tuned for supercritical water is desired for better simulation results.

TUNING PENG-ROBINSON EOS

For PR-EoS, the fugacity of a pure component can be evaluated from

$$\ln\frac{f}{P} = Z - 1 - \ln(Z - B) - \frac{A}{2\sqrt{2}B}\ln\left(\frac{Z + (1+\sqrt{2})B}{Z + (1-\sqrt{2})B}\right) \tag{12}$$

where f denotes fugacity, Z, compressibility factor,

$$A = aP / R^2T^2 \tag{13}$$

$$B = bP / RT \tag{14}$$

, and for PT-EoS, from

$$\ln\frac{f}{P} = Z - 1 - \ln(Z-B) - \frac{a}{2RTN}\ln\left(\frac{Z+M}{Z+Q}\right) \quad (15)$$

where

$$N = \left[bc + \frac{(b+c)^2}{2}\right]^{1/2} \quad (16)$$

$$M = (\frac{b+c}{2} - N)P/RT \quad (17)$$

and

$$Q = (\frac{b+c}{2} + N)P/RT \quad (18)$$

For each experimental data along the saturation line, applying the phase equilibrium condition

$$f^V = f^L \quad (19)$$

gives, from Eq. (12)

$$Z^V - \ln(Z^V - B) - \frac{A}{2\sqrt{2}B}\ln\left(\frac{Z^V + (1+\sqrt{2})B}{Z^V + (1-\sqrt{2})B}\right) = Z^L - \ln(Z^L - B) - \frac{A}{2\sqrt{2}B}\ln\left(\frac{Z^L + (1+\sqrt{2})B}{Z^L + (1-\sqrt{2})B}\right) \quad (20)$$

Hence a single value of *A* can be determined for the data point, and a value of α can be obtained through Eqs. (4) and (13). Similarly, a value of α can be calculated for PT-EoS using Eqs. (8), (15) and (19) for each saturated data point. For supercritical region or where a single vapor phase exists, fugacity for experimental data can be evaluated via

$$RT\ln\frac{f}{f^0} = (h-h^0) - T(s-s^0) \quad (21)$$

where h and s denote specific enthalpy and entropy, respectively, of the vapor at T and P, and the superscripts symbolize the thermodynamic property of a vapor at the same T and a low pressure, low enough for the vapor to be regarded as ideal gas. Then a value of α for PR-EoS can be obtained from Eqs. (4), (12) and (13) for each experimental data point. By the same way, a value of α for PT-EoS can be determined for each experimental data with Eqs. (21), (8), (15) and (13).

The value of α from experimental data for PR-EoS is plotted as symbols against Tr, the reduced temperature, in Figure 7. For Tr < 1, saturated steam data were used, and for Tr > 1, superheated steam at 400, 500 and 600°C under 30 MPa were employed. Also shown in Figure 7 are values of α calculated from Eqs. (5) and (6), using the real value of acentric factor, ω = 0.344, as a solid line, and using a refitted ω = 0.5491, as a dashed line. As can be seen, the dashed line fits the experimental data much better extending from Tr = 0.7 to 1.3. Therefore, A better performance is expected from PR-EoS with ω = 0.5491. The procedure of obtaining the new fitting is delineated as follows.

The value of α is re-plotted as $\alpha^{0.5}$ against 1-$Tr^{0.5}$ in Figure 8 for Tr between 0.85 and

1.35, a typical subcritical and supercritical region. The data exhibit a trend of straight line and are fitted by means of least square with the following line

$$\sqrt{\alpha} = 1 + 1.1401(1 - \sqrt{Tr}) \tag{22}$$

Comparing Eq. (22) with Eq. (5) gives $\kappa = 1.1401$, which is then plugged into Equation (6) to obtain a fitting value of ω = 0.5491.

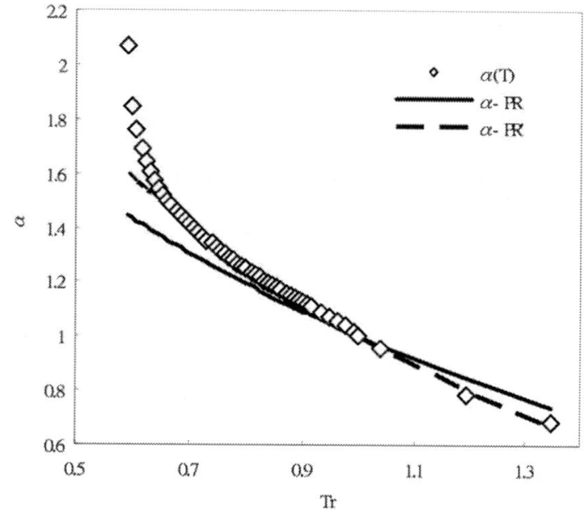

Figure 7. Comparison of α for PR-EoS.

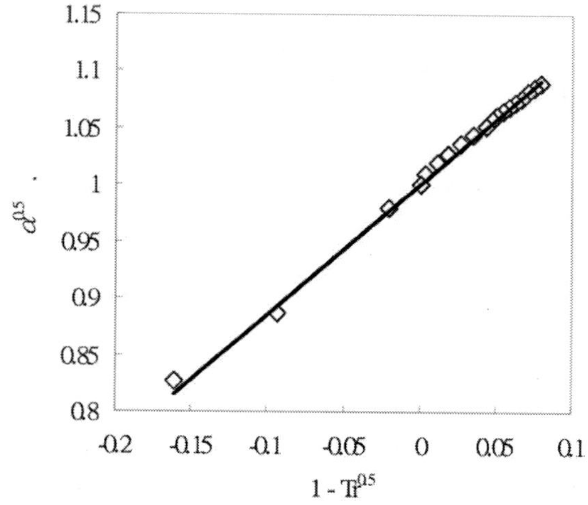

Figure 8. Fitting of α for PR-EoS.

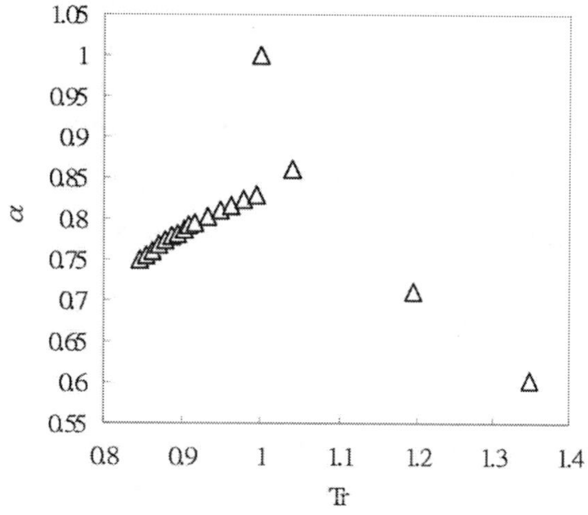

Figure 9. Values of α for PT-EoS.

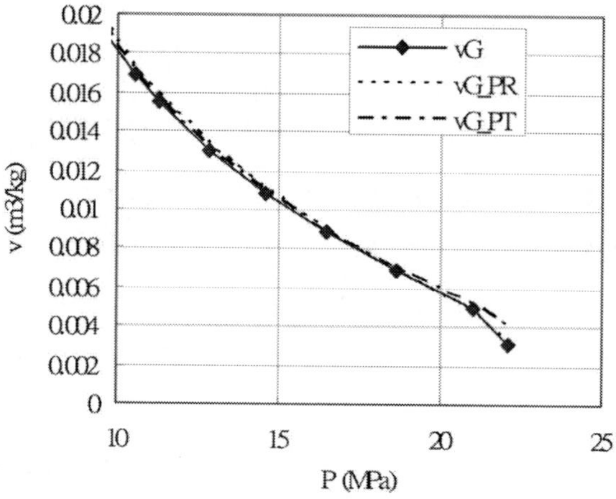

Figure 10. Specific volumes along the water saturation curve: retuned PR-EoS.

The values of α for PT-EoS from the same experimental data used for PR-EoS are displayed in Figure 9 against the reduced temperature. As can be seen, there is an abrupt jump in α value and a break in slope near the critical point, making retuning the PT-EoS extremely troublesome. Therefore the tuning of PT-EoS is not yet completed.

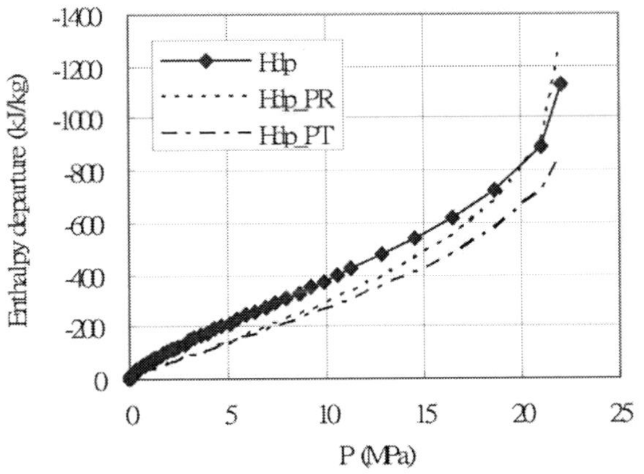

Figure 11. Enthalpy departures along the water saturation curve: retuned PR-EoS.

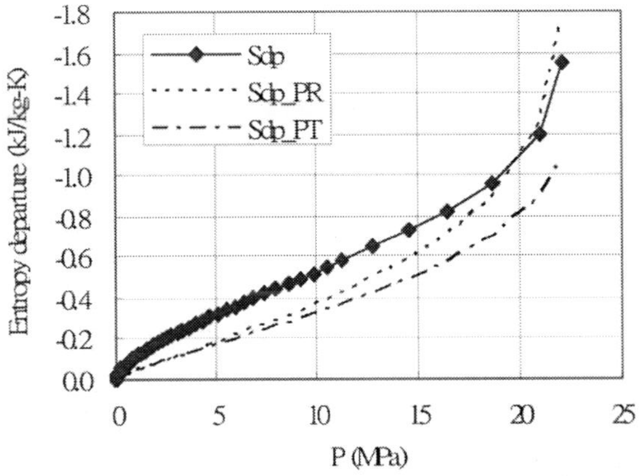

Figure 12. Entropy departures along the water saturation curve: retuned PR-EoS.

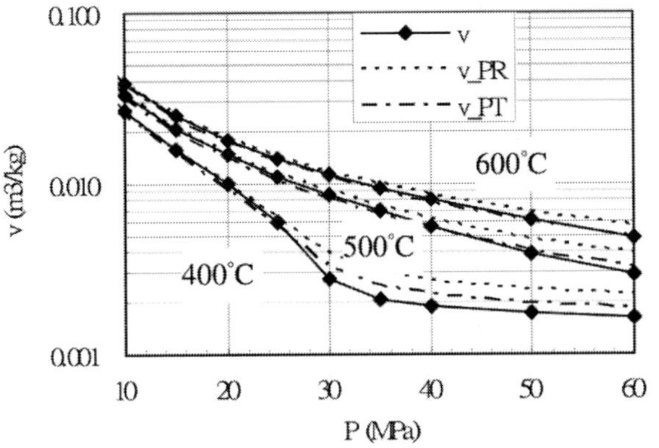

Figure 13. Specific volumes for the superheated vapor region: retuned PR-EoS.

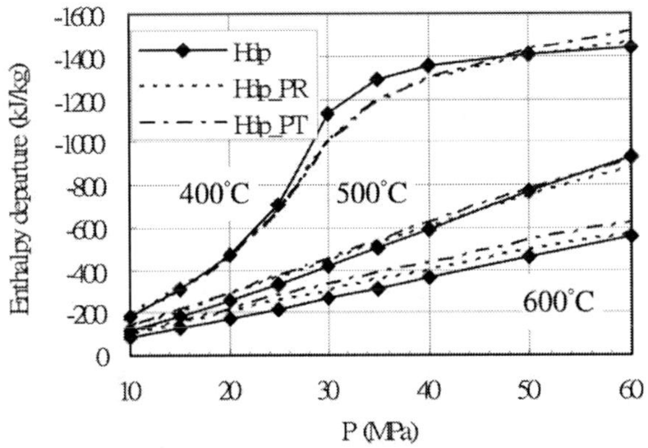

Figure 14. Enthalpy departures for the superheated vapor region: retuned PR-EoS.

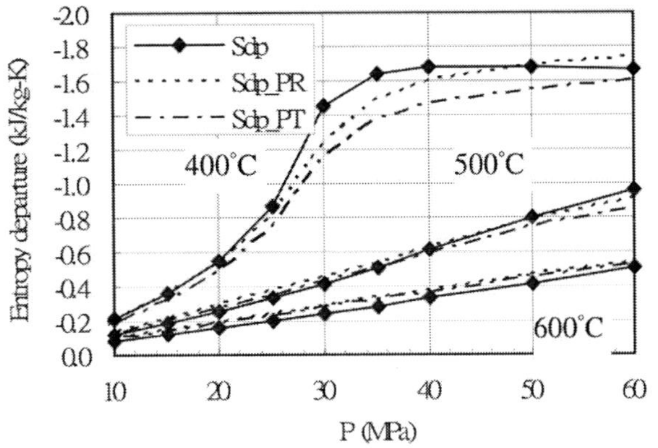

Figure 15. Entropy departures for the superheated vapor region: retuned PR-EoS.

The specific volumes, enthalpy departures and entropy departures are recalculated using PR-EoS with the new fitting ω = 0.5491, and presented in Figures 10 through 12 for the saturation curve and in Figures 13 through 15 for the typical supercritical region. For the saturated region, the prediction of the specific volume has improved dramatically, while the prediction of the enthalpy departure and the entropy departure has improved significantly, especially near the critical area. For the supercritical region, it is seen the prediction of enthalpy departure and entropy departure is better improved, and the performance of PR-EoS has generally exceeds that of PT-EoS, except for the specific volume.

ACKNOWLEDGMENTS

It is grateful for the financial support from I-Shou University (ISU95- 02-11).

REFERENCES

1. Watanabe, M., Inomata, H., Osada, M., Sato, T., Adschiri, T., and Arai, K., *Fuel*, **82**, 5, 2003, 545-552
2. Kim, Y. L., Chung, S. J., Kim, J. D., Lim, J. S., Lee, Y. W. and Yi, S. C., *Reaction Kinetics and Catalysis Letters*, **77**, 1, 2002, 35-42
3. Park, Y. K., Reaves, J. T., Curtis, C. W. and Roberts, C. B., *Journal of Elastomers and Plastics*, **31**, 2, 1999, 162-179
4. Tang, H. and Kitagawa, K., *Chemical Engineering Journal*, **106**, 2005, 261-267
5. Peng, D. Y. and Robinson, D. B., *Ind. & Eng. Chem. Fund.*, **15**, 1, 1976, 59-64
6. Patel, N. C. and Teja, A. S., *Chem. Eng. Sci.*, **37**, 3, 1982, 463-473
7. Redlich, O. and Kwong, J. N. S., *Chem. Rev.*, **44**, 1949, 233-244

On the Derivation of Damped Wave Difusion and Relaxation Equation from Stokes-Einstein Theory

Dr. Kal Renganathan Sharma PE
Adjunct Professor, Department of Chemical Engineering (MC110)
University of Illinois at Chicago, 202 Chemical Engineering Building
810 South Clinton Avenue, Chicago, IL 60607
Tel: (312) 996 3424

BACKGROUND

The study of transient diffusion has a variety of applications such as in semiconductor device manufacturing, active solid state processes, metal treatment, coating processes, metal fusion bonding, superconducting technology, etching a substrate, optical waveguides, gas separation processes, bleaching, dyeing and chemical modification of textiles and fiber, electrolysis, simultaneous diffusion and reaction catalytic, heterogeneous and homogeneous reactions, both traditional and novel separation methods, transport in living cells, oxygen and solute transfer in tissues, dispersal of pollutants etc.

Onsager (1931) argued that the Fourier's law of heat conduction (1822) and by analogy the Fick's law of diffusion contradicts the principle of microscopic reversibility. Fick's law is only an approximation of the description of transient diffusion and neglects the time needed for the acceleration of mass flow. The time needed for acceleration may be small and could be of the same order of magnitude of that of the collision time between molecules. In order to remove the paradox in the Fick diffusion model assuming an infinite speed of propagation, the damped wave diffusion and relaxation equation can be written by analogy with that suggested by Maxwell (1867), Morse and Feshbach(1953) and postulated independently by Cattaneo (1958) and Vernotte (1958) for heat conduction. Reviews on heat waves have been presented by Joseph and Preziosi (1989,1990) and Osizik and Tzou (1992). Very little work has been reported on the connection between macroscopic laws and miscroscale phenomena and the manifestation of the miscroscale phenomena in the macroscale. A comprehensive insight into the analytical solutions using the manifestations of the generalized transport equation is given in Sharma (2005].

The Cattaneo and Vernotte equation was found to be admissible within the framework of the second law of thermodynamics (Tzou, 1997). The relaxation time was suggested to be viewed as the time-lag between the heat flux vector and the temperature gradient when the response time is short. Some investigators have related the relaxation time to the electron-phonon collisions and the volumetric heat capacities of the electrons and the metal lattice (Ozisik and Tzou, 1994). Nernst (1917) suggested that at low temperatures in good thermal conductors heat may have sufficient "inertia" to give rise to oscillatory discharge. Landau (1941) found two speeds, one for ordinary sound, one for a second sound, which describe propagation waves of temperature. In Landau's theory there is no damping or dissipation and both sound speeds are associated with wave equations rather than telegraph equations.

Propagation of Landau waves in Helium II is specifically a quantum phenomenon. Other forms of non-Fourier heat conduction is the ballistic transport equations due to Chen (2001), dual phase lag model by Tzou (1997), Jeffrey (1924) and the EPRT, equation of phonon transport proposed by Majumdar (1993) and the microscopic two-step model suggested by Qiu and Tien (1992). An infinite order PDE was presented by Sharma (2003) using Taylor series expansion to fully describe the transient events. Experimentally measured values of relaxation times were reported in the order of a few microseconds by Tzou (1997) for steel at 400 ^{0}C. Mitra, Kumar et. al. (1995) and Kaminski (1990) have measured relaxation times for materials with a non-homogeneous inner structure of the order of 15-20 seconds.

The motivation for seeking a generalized law for mass diffusion where Fick's law becomes the particular case is six fold. The second is some of the singularities found in the description of transient mass diffusion using the parabolic diffusion equations. The first being the theory of Onsager. When transient concentration events are described using the Fick's law a "blow-up" occurs during short contact times in the expression for surface flux. This is so, for the cases of (Sharma, 2005); a) surface flux expression in a semi-infinite body subject to a step change in one of the boundary concentration, b) surface flux for a finite slab subject to constant wall concentration on either of its edges, c) concentration term in the constant wall flux problem in cylindrical coordinates in a semi-infinite medium solved for by the Boltzmann transformation leading to a solution in exponential integral d) in the short time limit, the solution from the parabolic diffusion equations for a semi-infinite sphere using the similarity transformation. The third reason for seeking alternate forms to Fick's law of mass diffusion is because light is the speediest of velocities. On examining the solution for the transient concentration, Landau and Lifshitz (1941) noted that for times greater than zero the concentration is finite at all points in the infinite medium except at infinite location . It can be inferred that the mass pulse has traveled at infinite speed. This is in conflict with the light speed barrier stated by the theory of relativity of Einstein. The fact that any speed of a moving object, including the thermal wave, must be less than the speed of light was examined by Kelly (1968) for diffusion. The fourth reason is the realization of the empirical nature of the development of the Fick's law of mass diffusion from observations at steady state. Use of it in the transient state is an extrapolation. The fifth is the over prediction of theory to experiment in important industrial process chromatography, adsorption, gel electrophoresis, restriction mapping, during manufacture of semi-conductor devices and drug delivery systems. A sixth reason is that the Flick's law breaks down also at small scales (Bejan (1988), Casimir (1938)). In this limit the flux is described by an expression similar to the one used in radiation heat transfer (Swartz and Pohl, 1989). The mass transport, for example, in dielectric crystalline materials is believed to be primarily by atomic or crystal vibrations. These vibrations travel as waves and the energy of the waves quantitated is the phonon, (Kittel, 1986).

DERIVATION OF GENERALIZED FICK'S LAW OF DIFFUSION FROM STOKES EINSTEIN FORMULATION

The Fick's law of diffusion can be written as;

$$J = -D\partial C/\partial x \qquad (1)$$

The generalized Fick's law of diffusion can be written as;

$$J = -D\partial C/\partial x - \tau_r \partial J/\partial t \qquad (2)$$

Eq. (2) is written by analogy with the generalized Fourier's law of heat conduction and relaxation (Sharma, 2005, 2006).

The Stokes-Einstein equation can be used to calculate diffusion coefficients in liquids:

$$D = k_B T/f = k_B T/6\pi\mu R_o \qquad (3)$$

Eq. (1) can be derived as follows. A rigid solute sphere is assumed for the molecule diffusing in a common solvent. The frictional drag force acting on the molecule opposing its motion is proportional to the velocity of the sphere;

$$\text{Drag force} = f\, v_1 \qquad (4)$$

Where f is the friction coefficient. From Stokes law (1850) for a sphere moving in a fluid, $f = 6\pi\mu R_o$. The driving force was taken by Einstein (1905) to be the negative of the chemical potential gradient defined per molecule.

$$-\nabla\mu_1 = (6\pi\mu R_o)v_1 \qquad (5)$$

Eq. (3) is valid when the molecule reaches a steady state velocity. This is when the net force acting on the molecule is zero. Solution is assumed to be ideal and dilute.

$$\mu_1 = \mu_1^0 + k_B T \ln(x_1) = \mu_1^0 + k_B T \ln c_1 - k_B T \ln c_2 \qquad (6)$$

For dilute solutions, c_2 far exceeds the solute concentration and can be taken as constant. The gradient at constant temperature, then is;

$$\nabla\mu_1 = k_B T\ \nabla c_1/c_1 = -(6\pi\mu R_o)v_1 \qquad (7)$$

$$\text{or } -k_B T/(6\pi\mu R_o)\ \nabla c_1 = c_1 v_1 = J'' \qquad (8)$$

Comparing Eq. (6) with the Fick's law of diffusion given in Eq. (1), the Stokes-Einstein relationship of Eq. (1) results. It can be seen that Eq. (8) is valid only at steady state. Often times, in transient applications a sudden step-change in concentration, i.e., the driving force is imposed on the system. The molecule will experience a accelerating regime prior to reaching steady state. During the accelerating regime;

$$-\nabla\mu_1 - (6\pi\mu R_o)v_1 = m dv_1/dt \qquad (9)$$

$$\text{or } k_B T \nabla c_1 + m c_1 dv_1/dt = -(6\pi\mu R_o)c_1 v_1 \qquad (10)$$

$$\text{or } -(k_B T/6\pi\mu R_o)\nabla c_1 = m/6\pi\mu R_o \, \partial J''/\partial t + J'' \qquad (11)$$

Eq. (10) can be written from Eq. (11) given that a slowly moving sphere is considered (i.e., $v dc/dt = 0$). Further for 1 molecule the concentration is not going to vary in its vicinity to a considerable measure for the term $v dc/dt$ to be significant. Eq. (11) is the generalized Fick's law of diffusion that accounts for the acceleration regime of the molecule as well as the steady state regime. . The relaxation time is then given by;

$$\tau_r = m/6\pi\mu R_o = mD/k_B T \qquad (12)$$

In terms of P_{tot} the system pressure for ideal gas the relaxation time can be written as;

$$\tau_r = MD\rho/P \qquad (13)$$

The velocity of mass diffusion is given by;

$$v_m = (D/\tau_r)^{1/2} = (k_B T/m)^{1/2} \qquad (14)$$

Eq. (14) can be rewritten in terms of the molar gas constant and molecular weight as;

$$v_m = (D/\tau_r)^{1/2} = (RT/M)^{1/2} \qquad (15)$$

From the Maxwell distribution of molecular speeds the average velocity of the molecule is given by $(8RT/M)^{1/2}$. The velocity of mass diffusion is off by a factor of 8 from the average velocity of a molecule in random motion. The kinetic representation of pressure can be written after observing that a molecule moving in a cube of dimensions l with a velocity of v_x undergoes a momentum change of $2mv_x$ upon one collision with the wall. The number of collisions on the wall can be estimated by first calculating the time taken by the molecule to move the roundtrip from the wall after a collision to the opposite wall and return as $2l/v_x$. The number of collisions undergone by a molecule is $v_x/2l$. The rate of transfer of momentum to the surface from the molecular collisions is then, mv_x^2/l. The total force exerted by all the molecules colliding can be obtained b summing the contribution from each molecule and the pressure is obtained by dividing the sum by the area of the wall and is given by;

$$P_{tot} = m/l^3 (v_{x1}^2 + v_{x2}^2 + v_{x3}^2 + \ldots\ldots] \qquad (16)$$

Let N_m be the number of molecules in the system and n the number of molecules per unit volume. Then Eq. [(16) can be rewritten after multiplying the numerator and denominator by N_m:

$$P_{tot} = mn \langle v_x^2 \rangle = \rho \langle v_x^2 \rangle = 1/3\rho \langle v^2 \rangle \qquad (17)$$

As the molecules treated as particles move in random there is no preferred direction in the box. Hence $v^2 = v_x^2 + v_y^2 + v_z^2$. The square root of v^2 is called the root mean squared speed of the molecule and is an widely accepted average molecular speed. From the ideal gas law $P_{tot} = \rho RT/M$. Combining this with Eq. (15);

$$1/3 \langle v^2 \rangle = RT/M = A_N k_B T/M \quad (18)$$

Comparing Eqs. (18) and Eq. (15) it can be seen the velocity of mass is 1/3 of the root mean square velocity. This could be due to the one dimensional diffusion.

Sources, Further Reading and References

Bejan, A., 1988, *Advanced Engineering Thermodynamics,* John Wiley, New York.

Casimir, H. B. G., 1938, "Note on the Conduction of Heat in Crystals", *Physica,* 5, 495 -500.

Cattaneo, C., 1958, "A Form of Heat Conduction which Eliminates the Paradox of Instantaneous Propagation", Comptes Rendus, **247**, 431- 433.

Chen, G., 2001, "Ballistic Diffusive Conduction Equations", *Phys. Rev. Lett.,* **86,** 11, 2297-2300.

Fourier, J. B., 1955, *Theorie analytique de la chaleur,* English Translation by A. Freeman, 1955, Dover Publications, New York.

Jeffreys, H., 1924, *The Earth,* Cambridge University Press, UK.

Joseph, D. D., and Preziosi, L., 1989, " Heat Waves", *Reviews of Modern Physics,* **61**, 41-73.

Joseph, D. D., and Preziosi, L., 1990, "Addendum to Heat Waves" *Reviews of Modern Physics,* **62**, 375-391.

Kaminski, W., 1990, "Hyperbolic Heat Conduction Equation for Materials with a No homogeneous Inner Structure", *ASME J. of Heat Transfer,* **112**, 555-560.

Kelly, D. C., 1968, "Diffusion " A Retavistic Appraisal", *American J. Physics,* **36**, 1968, 585 - 591.

Kittel, C., 1986, *Introduction to Solid State Physics,* John Wiley, New York (1986).

Landau, L., 1941, " The Theory of Superfluidity of Helium II", *J. Phys.,* **5**, 71.

Landau, L., and Liftshitz, E. M., 1987, *Fluid Mechanics,* Pergamon, UK, 1987.

Maxwell, J. C., 1867, "On the Dynamical Theory of Gases", *Phil. Trans. Roy. Soc.,* **157**, 49.

Morse, P. M., and Feshbach., 1953, H., *Methods of Theoretical Physics,* McGraw Hill, New York.

Majumdar, A., 1991, " Microscale Heat Conduction in Dielectric Thin Films", *ASME Conference Paper, Heat Transfer Division,* **184**, , 33-42.

Mitra, K, Kumar, S., Vedavarz, A., and Moallemi, M. K., 1995, "Experimental Evidence of Hyperbolic Heat Conduction in Processed Meat", *ASME J of Heat Transfer,* **117**, 568-573.

Nernst, W., 1917, *Die Theoretischen Grundalgen des n Warmestazes,* Knapp Halle.

Qiu, T. Q. and Tien, C. L., 1992, " Short-Pulse Laser Heating on Metals", *Int. J. Heat and Mass Transfer,* **35**, 719-726.

Onsager, L, 1931, "Reciprocal Relations in Irreversible Processes", Phys. Review, **37**, 1931, 405-426.

Ozisik, M. N., and Tzou, 1994, D. Y., "On the Wave Theory of Heat Conduction", *ASME J of Heat Transfer,* **116**, 526- 535.

Ozisik, M. N. and Tzou, D. Y., 1992, *On the Wave Theory in Heat Conduction,* ASME Winter Annual Meeting, Anaheim, CA..

Sharma, K. R., 2005, *Damped Wave Transport and Relaxation,* Elsevier, Amsterdam.

Sharma, K. R., 2006, "Manifestation of Acceleration in Transient Heat Conduction", J of Thermophysics and Heat Transfer, 26, 4, 799-808.

Sharma, K. R., 2003, Surface *Renewal Theory for Mass Transfer Coefficient using Infinite Order PDE Inorder to Capture non-Fickian Diffusion,* 225th ACS National Meeting, New Orleans, LA.

Sharma, K. R., and Turton, R., 1998, "Mesoscopic Approach to Correlate to Surface Heat Transfer Coefficients using Pressure Fluctuations in Dense Gas-Solid Fluidized Beds", *Powder Technology,* **99**, 109-118.

Swartz, E. T., and Pohl R. O., 1989, "Thermal Boundary Resistance" *Reviews of Modern Physics,* **61**, 605 - 668.

Tzou, D. Y., 1993, "An Engineering Assessment to the Relaxation Time in Thermal Wave Propagation", *Int. J. Heat Mass Transfer,* **36**, 7, 1845-1850.

Tzou, D. Y., 1997, *Macro-to Micro Scale Heat Transfer: The Lagging Behavior,* Taylor and Francis, Washington, DC.

Vernotte, P., 1958, "Les Paradoxes de la Theorie Continue de l'equation de la Chaleur", *C. R.. Hebd. Seanc. Acad. Sci. Paris,* 246, **22**, 1958, 3154 - 3155.

Eight Ensembles in Treatment of Statistical Thermodynamics Classical Systems

Tahir Cagin, Chemical Engineering, Texas A&M University, College Station, TX 77843

The conventional statistical treatment of classical fluids usually focuses on isothermal ensembles in studying the thermodynamic of these systems. In this talk, I will outline the classfication of classical thermodynamics systems and their interrelationship by presenting eight different statistical thermodynamic ensembles. I will also address thermodynamics observables including response functions and their relationships to the microscopic variables defining molecular level driving forces in physico-chemical processes.

Sublimation Pressure from Solubility Data of Solids in Supercritical Solvents

Claudio A. Faúndez[1] and José O. Valderrama[* 2,3]
(1) Depto. Física, Fac.de Física y Mat., Univ. de Concepción, Casilla 160-C, Concepción-Chile
(2) Depto. Ing. Mecánica, Fac. de Ingeniería, Univ. de La Serena, Casilla 554, La Serena-Chile
(3) Centro de Información Tecnológica, Casilla 724, La Serena – Chile

Abstract

Sublimation pressure from solubility data of solids in different supercritical solvents is determined and solvent effects are discussed. Tree binary gas-solid systems containing anthracene, napthalene and phenanthrene as solutes and carbon dioxide, ethane, fluoroform as solvents are considered in the study. The Peng-Robinson equation of state with the mixing rules proposed by Wong and Sandler are used to evaluate the fugacity coefficient ϕ^s that appears in the classical solubility equation which relates the mole fraction of a dissolved solid in a compressed gas phase: $y=(P^{sub}/P\phi^s)[\exp\{(V^s(P-P^{sub})/RT\}]$. The sublimation pressure P^{sub} of the solid is considered as a parameter to be determined by regression analysis of experimental solubility data (TPy), and theoretically it should be independent of the solvent. Thus, an optimization problem, in which the difference between correlated and experimental solubility is to be minimize, is solved using a method based on genetic algorithms. The results show that the determination of P^{sub} using high pressure solubility data is reliable and produce unique values of sublimation pressure independent of the solvent present in the mixtures.

Introduction

Phase equilibrium calculations of a solid dissolved in a compressed gas, at a pressure P and a temperature T can be performed using the fundamental equation of phase equilibria which leads to a simple equation that relates the solubility y the pressure P, and the temperature T. The equation also contains other properties such as the sublimation pressure P^{sub}, the molar volume of the solid V^s and the fugacity coefficient of the solid component in the high pressure gas ϕ^s [1]. Of all these properties involved in the calculation of the solubility of the solid in the high pressure gas, the sublimation pressure has received low attention in the literature, although it is directly related to the solubility. The molar volume does not have a strong influence on the calculations and the fugacity coefficient ϕ^s can be estimated from an appropriate equation of state and mixing rules. The sublimation pressure is usually small for common industrially important solids and in many cases available experimental techniques cannot be used to obtain accurate values.

In theory, the calculated sublimation pressures should be independent of the solvent used, but this statement must be proved. In this article the influence of the type of solvent in which the solid is dissolved on the calculation of P^{sub}, is analyzed. For this, the sublimation pressure of anthracene, naphthalene and phenanthrene in solvents of different nature (carbon dioxide, ethane and fluoroform), are determined to prove what in theory seems to be clear.

Solubility Calculations

The theory of solid solubility in a compressed gas is found in standard books [1, 2]. If subscript 2 stand for the solid component in the gas-solid mixture, the so-called solubility equation is:

$$y_2 = \frac{P_2^{sub}\, e^{\frac{V_2^{sol}}{RT}(P - P_2^{sub})}}{P\varphi_2} \tag{1}$$

Here, P_2^{sub} is the sublimation pressure of the pure solid, V_2^{sol} is the solid molar volume, all at the temperature T and φ_2 is the fugacity coefficient of solid (2) at the pressure P. The fugacity coefficient is calculated from standard thermodynamic relations [2].

The Peng-Robinson equation (PR) [3], with the mixing rules proposed by Wong and Sandler [4], are used as the thermodynamic model to evaluate the fugacity coefficient φ_2. The excess Gibbs free energy that appears in the Wong-Sandler mixing rule (WS) is represented by the van Laar model (VL). Therefore, the model contains three parameters (the van Laar constants A_{12} and A_{21} and the k_{12} parameter included in the mixing rule). This model is designed here as PR/WS/VL and the expressions for the fugacity coefficient can be found elsewhere [1]. The problem is then reduced here to determine the three parameters of the thermodynamic model and the sublimation pressure P_2^{sub} that appears in the solubility equation (1), using available high pressure TPy data for gas-solid systems.

Binary system containing anthracene, napthalene and phenanthrene as solute in various solvents (carbon dioxide, ethane, fluoroform) were considered in the study. The data consisted of 30 isotherms and a total of 381 PTy data points.

Results and Discussion

The accuracy of the model and the method used to calculate the sublimation pressure is evaluated by considering the deviations between experimental [5-8] and calculated values of the solubility of the solid in the high pressure gas. Literature information indicate that deviations are less than 10% using the PR/WS/VL model [9], deviations that were found of the same order in this study. The Table 1 shows the results obtained using PR/WS/VL model for gas-solid systems considered in this study. The calculated sublimation pressure and the deviations with respect to experimental data are given. For systems which contain naphthalene as solute, the relative deviations vary from -0.22% to 7.5%. For the systems which contain phenanthrene as solute, the relative deviations vary from 0.09% to -6.8% and for systems which contain anthrace as solute, the relative deviations vary from -0.05 % to 7.7%.

Fig. 1 shows sublimation pressure calculated from napthalene for all isotherms with different solvents. The symbols in the Fig. are (*) CO_2 for T= 308, 328, 333 and 338 K; and (Δ) ethane for T= 318 K. The values for fluoroform are not shown in the figure since the isotherms and the calculated values coincide with those of the system naphthalene + CO_2 and naphthalene + ethane. In the figure, sublimation pressure is in bar and T in Kelvin. The results shown in Fig. 1 indicate that the proposed method is reliable enough to estimate the sublimation pressure of solids using solubility data of the solid in high pressure gases. The solvent present in the mixture shows to be irrelevant in determining P^{sub} as expected.

Table 1: Sublimation pressures of the solid solutes at all temperatures studied and relative deviations between calculated and experimental values. N is the number of experimental data.

Naphthalene+	T(K)	N	$P^{sub}x10^4$(bar)	% ΔP^{sub}
CO_2	308	9	2.705	7.46
	328	16	16.060	-0.32
	333	19	24.070	1.62
	338	7	33.790	2.01
Ethane	308	15	2.892	1.06
	318	12	7.194	-2.39
	328	14	15.900	0.68
	308	6	3.090	-5.71
	318	8	7.089	-0.90
Fluoroform	308	6	3.026	-3.52
	318	6	7.099	-1.04
	328	6	16.930	-5.75
Phenanthrene+			$P^{sub}x10^6$(bar)	
CO_2	313	26	0.906	0.33
	323	30	2.677	-4.12
	333	30	6.947	-1.73
Ethane	318	6	1.546	-0.32
	328	6	4.249	-0.66
	333	7	6.993	-2.40
Fluoroform	318	4	1.646	-6.81
	328	4	4.276	-1.30
Anthracene+			$P^{sub}x10^7$ (bar)	
CO_2	323	8	1.89	7.76
	343	9	19.10	-2.59
	313	30	0.61	0.16
	323	30	2.05	-0.05
	333	30	6.38	0.08
Ethane	308	4	0.31	4.32
	323	10	2.07	-1.02
	343	7	18.60	0.10
Fluoroform	328	8	3.67	-0.60
	343	8	18.80	-0.98

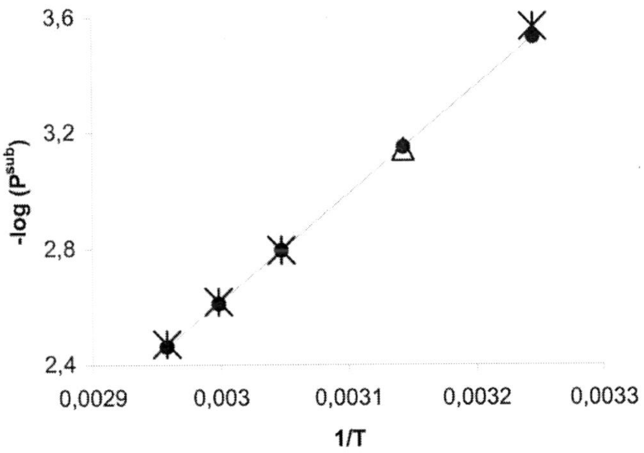

Fig. 1: Calculated sublimation pressure of Naphthalene.: (*) CO_2 for T= 308, 328, 333 and 338 K; and (Δ) ethane for T= 318 K. (•) experimental sublimation pressure.

Conclusions

Sublimation pressure of solid from high pressure solubility data are calculated using genetic algorithms. Based on the results and discussion presented in this study, the following main conclusions are obtained:

i) the low deviations between experimental and calculated values of the sublimation pressure show that the thermodynamic model PR/WS/VL is appropriate to model these systems and to estimate the sublimation pressure of solids.

ii) the genetic algorithms method has shown to be a good tool to solve the optimization problem studied here, providing accurate global optima.

Acknowledgements

The authors thanks the support of the National Commission for Scientific and Technological Research (CONICYT-Chile), through the research grants FONDECYT 1050410 and 1040285. JOV also thanks the Direction of Research of the University of La Serena-Chile for permanent support through several research grants and the Center for Technological Information (CIT, La Serena-Chile), for computer and library support. CAF thank the Dept. of Physics of the University of Concepción-Chile for special support that made possible the preparation of this work.

References

[1] J.M. Prausnitz, R.N. Lichtenthaler, E. Gomes de Azevedo, Molecular Thermodynamics of Fluid-Phase Equilibria. Prentice Hall International Series, New Jersey, USA, (1999).

[2] S.M. Walas, Phase Equilibria in Chemical Engineering, Butterworth Pub.-USA, (1985).

[3] D.Y. Peng, D.B. Robinson, Ind. Eng. Chem. Fundam., 15, 59-64 (1976).

[4] D.S. Wong, S.I. Sandler, AIChE Journal 38 671-680 (1992).

[5] K.P. Johnston, D.H. Ziger, C.A. Eckert, Ind. Eng. Chem. Fundam. 21, 191-197 (1982).

[6] G. Anitescu, L.L Tavlarides, J. Sup. Fluids 10, 175-189 (1997).

[7] W. J. Schmitt, R.C. Reid, Fluid Phase Equilib. 32, 77-99 (1986).

[8] M. McHugh, M.E. Paulaitis, J.Chem.Eng. Data, 25 326-329 (1980).

[9] J.O. Valderrama, J. Zavaleta, Ind. Eng. Chem. Res., 44, 4824-4833 (2005).

Using the Solvation Model to Predict the Salt Effect on Vapor-Liquid Equilibrium

Shuzo Ohe

Department of Chemical Engineering, Graduate School of Engineering,

Science University of Tokyo, 1-3 Kagurazaka, Shinjuku-ku, Tokyo 162-8601, Japan

ABSTRACT

The vapor-liquid equilibrium (VLE) of a mixture containing dissolved salt behaves quite differently from a mixture composed of volatile components alone. When, to a volatile component system in its equilibrium state, a salt is added, the composition of the vapor phase, as well as the temperature or pressure, will change to another equilibrium state. This change, called the salt effect on VLE, means that prediction requires a different approach than that for the no-salt system.

This is because the structure of the salt-containing system, or electrolyte system, is entirely different from that of the no-salt system, or non-electrolyte system. A typical model of the configuration of a non-electrolyte system is the local composition model. In contrast, a typical model of the configuration of an electrolyte system is the solvation model which we originally presented. The configuration of the non-electrolyte system is entirely different from that of the electrolyte system.

The formation of solvate in electrolyte solutions is well known. The solvation numbers for each ion of the salts have been reported by Marcus, who used Stoke's radii. We adopted Marcus's solvation numbers at infinite dilution and confirmed that the numbers determined by our model coincide with Marcus's ion solvation numbers. We have developed a prediction method that can express such a system thermodynamically based on the solvation model. Our solvation model can express the salt effect fully in terms of change of liquid phase composition and total vapor pressure.

Another approach is a modification of the local composition model, which was originally developed for non-electrolyte systems. Some methods using this approach include the Wilson and NRTL equations. This model extends the local composition model to deal with salt systems by simply adding electrolyte terms. Such equations have some drawbacks or deficiencies.

1. The modification is incongruous: A non-electrolyte model modified for electrolyte solutions is a misapplication, because of the difference in liquid configuration between non-electrolyte and electrolyte solutions.

2. The extended term is the same as the original term: We examined the extended "electrolyte" NRTL and can present evidence that it is in fact the same equation as in the original "non-electrolyte" NRTL. This means that the modified local composition equation is irrelevant. Furthermore, for most of the observed data we got the same result using the original NRTL as we got using the modified NRTL. Therefore, such equations cannot represent the system satisfactorily. They can only fit the data, by optimizing the error. Consequently, they cannot predict the equilibrium.

In our model, we assume that:

1. Both volatile components make solvates with each ion of a salt.
2. The salt is perfectly ionized to anion and cation.
3. Solvates cannot contribute to vapor liquid equilibria.
4. A free volatile component can contribute to the equilibria.
5. The free volatile component is not affected by the solvate.
6. The activity coefficient is divided into two parts:
    - expression of non-ideality
    - vapor pressure lowering

Thus, the number of liquid molecules that determine vapor-liquid equilibria changes from that in the original composition. (We call this resulting changed composition the "effective liquid composition".)

Using these assumptions, our model affords a more versatile method for predicting and correlating VLE for electrolyte solutions. In our model we have two methods for determining solvation numbers. The first method uses data on the elevation of the boiling point of a pure solvent with a salt. The second works directly from the observed data.

Our model can both predict the salt effect on vapor liquid equilibrium and correlate the observed data. We call the first method the predictive method and the second the correlative method. Using the predictive method with data on the boiling point elevation of a pure solvent with salt, we can predict and confirm the breaking of the azeotropic point (for example, in an ethanol, water, and NH_4Cl system at 101.3 kPa). At present, no other method can attain this result. We applied the solvation model to predict the methanol, ethanol, water and calcium chloride system, with quite satisfactory results. The solvation numbers used for each volatile component were those determined from the constituent binary volatile components of the salt system.

Our method has the following advantages:

1. The solvation number can be obtained from the vapor pressure lowering data of a pure solvent with salt, or from the salt effect data.
2. The ion solvation number reported independently can also be used.
3. We can predict VLE from the solvation numbers.

Moreover, our model expresses the salt effect more meaningfully. It can explain and demonstrate thermodynamically the mechanism of the salt effect on VLE. A further notable advantage of our model is that the visual interpretation of our parameters reflects more realistically the actual chemical structure of the liquid solution, making it easier to comprehend. We have applied our method for almost all systems reported in the literature with satisfactory results.

INTRODUCTION

It is well known that an azeotropic mixture cannot be separated by means of an ordinary distillation method and that some kinds of salt are significantly effective in reducing the azeotrope. For this reason, investigators in both academia and industry have studied the salt effect on vapor–liquid equlibria. It is quite important to clarify the mechanism of this effect and to advance the method of correlation and prediction.

Though many investigations have been performed in this area, most have been modifications of a conventional thermodynamic model for a non-electrolyte component system, such as the activity coefficients model. These models combined an extended non-electrolyte activity coefficients model and the Debye–Hückel term. The activity coefficients model term was used for describing short-range interaction by van der Waals force. The Debye–Hückel term was used for describing long-range interaction according to the Coulomb force.

Other examples include Tan's equation [Tan, 1987], which uses the Wilson equation to describe short-range interaction; the electrolyte NRTL model [Chen, 1982, 1986], which uses the NRTL equation; the extended UNIQUAC model [Sander 1986, Macedo 1990] and the LIQUAC model [Li 1994, Polka 1994] using the UNIQUAC equation; Kikic et al.'s method [Kikic 1991], which uses the UNIFAC equation; Achard et al.'s method [Achard 1994], which uses a modification of the UNIFAC equation; and Yan et al.'s method [Yan *et al.* 1999], which applies the LIQUAC model to the atomic group contribution method.

However, most of the mixtures in the salt effect system are composed of solvents with salt. Thus, a need exists for a model that is capable of explaining contributions of these two different kinds of components. We originally presented a solvation model in 1975. I theorized that alteration in vapor phase composition might be caused by the formation of solvates by interaction between solvents and salt. I also thought that the solvates might play a part in lowering the vapor pressure of the system. First I proposed, to predict the salt effect on vapor–liquid equilibrium, a model in which a salt forms a preferential solvate with one of the solvents in a binary solvent system [Ohe 1976]. This model was successfully applied to correlate and express the salt effect for 386 binary solvents 47 single-salt systems [Ohe 1991].

This first model, however, has a drawback, in that it cannot sufficiently express the lowering of the vapor pressure, since the lowering is observed over the entire range of the liquid composition of a solvent. Based on the model, the lowering is zero at the end of the composition of the solvent, which does not match the experimental fact.

I then modified the model to overcome this drawback. This subsequent model employed, instead of a preferential solvation, a solvation between the salt of each of solvent in the mixture. Alteration of vapor phase composition by the salt effect is thought to be caused by the difference, between solvents, in their degree of solvation. This difference of degree corresponds to the preferential solvation [Ohe 1998, 2000]. This latter work examines the relation between solvation number and correlation error. The relation is used to determine the solvation number. And the solvation numbers were used to predict the salt effect on vapor liquid equilibria.

ACM MODEL

To date, extended models for electrolyte solution have been presented for almost all ACMs. Among them, one of the most widely used is that presented by Mock and Chen in 1986. They extended the original NRTL of the Local Composition Model for non-electrolyte solutions, for use with electrolyte solutions. This is the extended model for NRTL. Mock and Chen thought that each species had one of three orientations, with the cation at the center; with the anion; or with a molecule at the center.

In the equation of the electrolyte or extended NRTL model, the non-electrolyte term is the original NRTL. The electrolyte term is the extended NRTL. In the original case, three parameters were needed. In contrast, the electrolyte case requires nine parameters. However, I think this extension of the NRTL is not entirely reasonable. The original is based on the local composition model. It does not take into consideration strongly associated species such as are found in electrolyte solutions. In other words, it is not symmetrical.

Notice the following interesting result, in a methanol, water, and calcium chloride system. The fitting was done using only the original NRTL equation – not using the electrolyte NRTL. First, the original NRTL was applied to the mixture, using calcium chloride as the third component. Second, optimization was done by comparing the activity coefficients from the NRTL equation with those gained from observed data.

The fitting is almost perfect. Such a close fit is easy, when using such a large number of parameters – in this case, nine. But the number of parameters is not the point. The basis for accuracy is in the approach used. The point, then, is that the original NRTL can already provide a high degree of capability, even when applied to electrolyte solutions. The essential capabilities of the electrolyte NRTL are also possessed by the original NRTL. Therefore, any data that can be fit using the electrolyte NRTL can just as well be fit using the original NRTL.

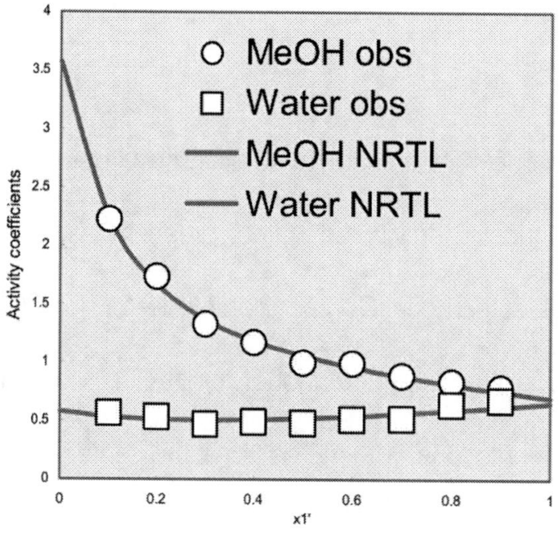

Figure 1. Methanol+Water+$CaCl_2$ (0.1 mole fraction) System at 101.3kPa

The fitting is perfect; however, there are problems with the resulting parameters. For example, the values of the alpha parameters vary greatly from those defined by Renon. Typical values for a volatile system should range from 0.1 to 0.5. But you can see here that alpha 31, alpha 23, and alpha 32 are 6.5624, 9.4757, and 9.5429, which are quite different from the originally defined values. This suggests two things: first, that the original NRTL possesses the capability to express the characteristics of the activity coefficient of an electrolyte system; and second, that the modified NRTL also has this capability

Table 1. NRTL Parameters

<table>
<tr><td colspan="2">$\tau_{12} = -0.3544$
$\tau_{21} = 1.286$
$\alpha_{12} = 0.3$</td></tr>
<tr><td>$\tau_{1,3} = 6.6682$
$\tau_{3,1} = -0.5072$
$\alpha_{1,3} = 0.2291$
$\alpha_{3,1} = 6.5624$</td><td>$\tau_{2,3} = 1.5556$
$\tau_{3,2} = -0.7556$
$\alpha_{2,3} = 9.4757$
$\alpha_{3,2} = 9.5429$</td></tr>
</table>

SOLVATION MODEL

Figure 2 shows two types of models. The first is the local composition model, on which the NRTL model is based. The second is the solvation model that we first presented in 1971. It is clear that in the electrolyte solution, the electrolyte associates strongly with the solvent molecules. Therefore, in the electrolyte solution, it is illogical to apply the local composition model.

We have been studying solvation since 1975. Our first model was the preferential solvation model. In 1996, we developed a solvation model in which each volatile component is solvated with each ion of the salt. We examined preferential solvation in specific cases, using a solvation model in which a non-affinitive component does not make a solvate with salt.

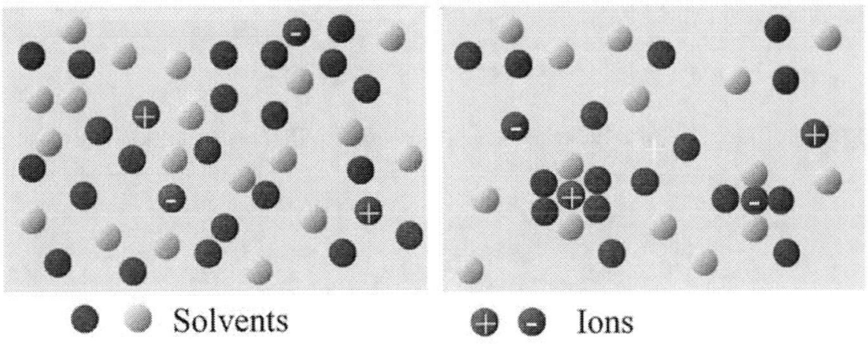

Local composition Solvation

Figure 2. A local composition and a solvation model

Molecules solvated with salt cannot evaporate, because the salt itself cannot evaporate. Only free molecules – not solvated molecules – can contribute to vapor-liquid equilibria. Thus, the number of liquid molecules that determine vapor-liquid equilibria changes from that in the original composition. We will call this resulting changed composition the "effective liquid composition".

We make the following assumptions:

1. Both volatile components make solvates with each ion of a salt.
2. The salt is perfectly ionized to anion and cation.
3. Solvates cannot contribute to vapor liquid equilibria.
4. A free volatile component can contribute to the equilibria.
5. The free volatile component is not affected by the solvate.
6. The activity coefficient is divided into two parts:
   - expression of non-ideality
   - vapor pressure lowering

1. SOLVATION NUMBER

The solvation number is assumed to be the sum of each ion solvation number. The ion solvation number is determined from Stokes' radii independently from the salt effect on VLE. The reported ion solvation number by Marcus is at infinite dilution. The ion solvation number can be expressed as

$$Sio_{\text{ion (solvent)}} = So_{\text{ion (solvent)}} - k \cdot x_{\text{ion}}$$

from the examination using salt effect data as shown in Fig. 3. The solvation number $Sio_{\text{salt (solvent)}}$ is given by

$$Sio_{\text{salt (solvent)}} = \sum Sio_{\text{ion (solvent)}},$$

considering the ionic valency. For example, $Sio_{CaCl2(H2O)}$ for $CaCl_2$-water is calculated as

$$Sio_{CaCl2(H2O)} = Sio_{Ca2+(H2O)} + 2\, Sio_{Cl-(H2O)}.$$

The values at the $x_{\text{ion}} = 0$ are the ion solvation numbers reported by Marcus.

In Fig. 3, each line is determined from the salt effect data. The ion solvation number can be expressed as a straight line. The number is decreased by the increase in ion concentration. The intercept shows the ion solvation number at infinite dilution. The values coincide approximately with the original ion solvation number of Stokes' radii.

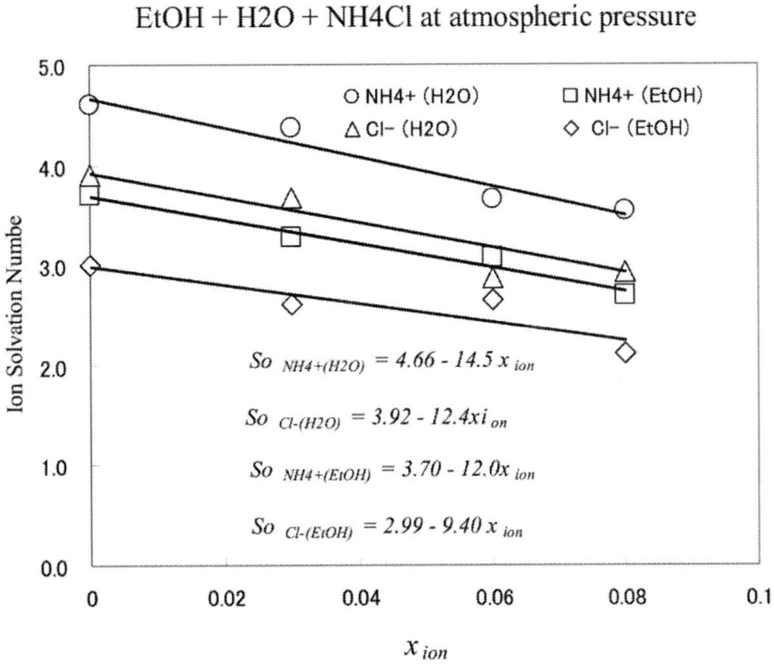

Figure 3. Ion Solvation Number and Ion Concentration

2. LOWERING OF VAPOR PRESSURE

Dissolved salt, which is non-volatile substance, lowers vapor pressure at a given temperature of a solvent. When the behavior of a single solvent and a salt system is assumed to be non-ideal and complete ionization of salt, the pressure of the system is given by Eq. (1):

$$\pi = P_{solvent} \times a_{solvent} \tag{1}$$

The activity is a product of activity coefficient and mole fraction of solvents: $a_{solvent} = \gamma_{solvent} \times x_{solvent}$. Thus the activity coefficient for the pressure is determined from Eq. (1) as follows:

$$\gamma_{solvent} = \pi / (P_{solvent} \times x_{solvent}) \tag{2}$$

The model assumed salt showed complete disassociation. It is assumed that lowering of vapor pressure of pure solvent system is caused by solvated molecule, which does not contribute to vaporization. The activity of the solvent becomes

$$a_{solvent} = \frac{x_{solvent} - S_0 \cdot x_{salt}}{1 - S_0 \cdot x_{salt}} \tag{3}$$

Then for the solvation number, we can get

$$S_0 = \frac{1 - x_{salt}}{x_{salt}} \cdot \frac{1 - \gamma_{solvent}}{1 - \gamma_{solvent} \cdot x_{solvent}} \tag{4}$$

3. EFFECTIVE COMPOSITION OF SOLVENTS SYSTEM CONTAINING SALT

We can assume that the concentration of the solvent is decreased according to the number of solvated molecules. Thus, the actual solvent composition participating in the vapor–liquid equilibrium is changed. Assuming that a salt forms the solvates with each component and that the solvated molecules do not contribute to vapor–liquid equilibria, the effective mole fraction $x_{ia}^{'}$ for the i-th solvent, on a salt free basis, can be obtained using Eq. (5). Since the sum of solvent mole fraction is equal to 1 - x_{salt} , and if the solvation number for the i-th solvent is calculated as $S_{i0} \cdot x_i^{'}$, then Eq. (5) can be obtained as follows:

$$x_{ia}^{'} = \frac{x_i - S_{i0} \cdot x_i^{'} \cdot x_{salt}}{1 - x_{salt} - \sum_{k=1} S_{i0} \cdot x_k^{'} \cdot x_{salt}} \tag{5}$$

We determined experimentally that the solvation number *S* is expressed by the relation between a single solvent S_0 and its composition $x_i^{'}$. There is a linear relationship between the solvation number S and the composition of the solvent $x_i^{'}$.

$$S_i = S_{i0} \cdot x_i^{'}$$

This relation stems from the molecular numbers of the solvents for each salt molecule changing stoichiometrically as the liquid composition changes. This, then, is one of the strongest pieces of evidence in support of the solvation model. Furthermore, the relation has a physical meaning. It is that the solvation number *S* is dependent upon the solvent concentration remaining in a state of being capable of making solvates of salt that are constant at a given mole fraction concentration, since the number of solvent molecules that can be solvated increases according to the amount of solvent.

4. PREDICTION OF SALT EFFECT FROM SOLVATION NUMBER

In the case of a mixed solvents system, lowering of the vapor pressure may be treated in a similar manner. As regards the total pressure of a solvents system, Eq. (6) can be used for a non-ideal solution, corresponding to Eq. (1) for the pure solvent system.

$$P_{solvent} = \sum P_i \gamma_i' \cdot x_{ia}' \qquad (6)$$

The activity coefficient for vapor pressure lowering is assumed to be the average of the activity coefficient of each solvent with its mole fraction. Eq. (7) may be derived from the above assumption.

$$\gamma_{mix,solvent} = \sum \gamma_{i,solvent} \cdot x_i' \qquad (7)$$

Therefore, the basic equation for calculating the salt effect on vapor–liquid equilibrium, corresponding to Eq. (1), is Eq. (8).

$$\pi = P_{solvent} \gamma_{mix,\ solvent} (1 - x_{salt}) \qquad (8)$$

Since $P_{solvent}$ is determined from Eq. (6), substitution of Eq. (6) into Eq. (8) gives the relation

$$\pi = (\sum P_i \gamma_i' \cdot x_{ia}') \gamma_{mix,solvent} (1 - x_{salt}) \qquad (9)$$

The relation between the activity coefficient as conventionally defined and that given by Eq. (9) is

$$\gamma_i = \gamma_i' \cdot x_{ia}' \cdot \gamma_{mix,solvent} (1 - x_{salt}) / x_i' \qquad (10)$$

5. DETERMINATION OF SOLVATION NUMBER

Our model's ion solvation numbers derived from Stokes' radii are the values at infinite dilution. Thus we can predict two facts: the solvation numbers will be at maximum value; and, as the salt concentration increases, the solvation number can only decrease. Our examination of the observed data confirmed these facts, and our model shows these relations. This outcome qualitatively verified the consistency of our model's predictive capability.

There are two types of methods for determining solvation numbers. One determines directly from observed data of salt effect on VLE. The other uses data on the elevation of the boiling point of a pure solvent with a salt. Hence, our model can both correlate the observed data and predict the salt effect on vapor liquid equilibrium. Figure 4 shows a comparison of the results determined using the two methods. The difference is larger at the lowest salt concentration. Then it decreases as the salt concentration increases.

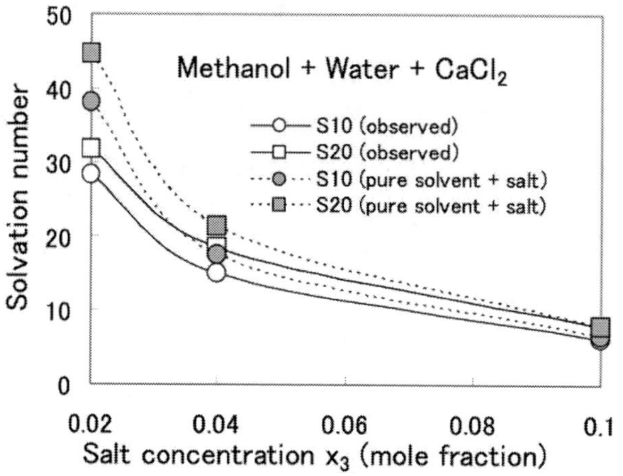

Figure 4. Comparison of Solvation Numbers Determined Using the Two Methods

The formation of solvate in electrolyte solutions is widely well known. The solvation numbers for each ion of the salts have been reported by Marcus, who used Stoke's radii. We adopted Marcus's solvation numbers and confirmed that the numbers determined by our model coincide with Marcus's numbers. Table 2 shows the solvation numbers derived from Marcus, and the results we calculated. The solvation number under the caption "solvation" was determined from observed data on the salt effect on VLE. The solvation number under the caption "Stoke's radii" was calculated from the solvation number for each ion at infinite dilution of salt, as reported by Marcus. The solvation number under the caption "pure solvent" was determined from observed data on vapor pressure lowering or boiling point raising in a solution of pure solvent with salt.

The solvation numbers for the system of Methanol (1) + Water (2) + NaNO3 (3) (8 mole %) are as follows.

	S_{10}	S_{20}	$S_{20} - S_{10}$
Solvation	6.23	9.24	3.01
Stoke's radii	8.0	9.8	1.8
Pure solvent	6.32	9.34	3.02

It is clear that the solvation number of "pure solvent" coincides with that of "solvation". This signifies the potential of our method to predict the salt effect on VLE. The difference between S20 and S10 is a decisive factor in calculating the salt effect, especially in the calculation of effective liquid composition. The value of the Stoke's difference shows the prediction capability qualitatively. The values shown for Stokes' radii are the greatest. This is because they are at infinite dilution. The values of definite concentration are less than those of Stokes' radii because of the activity of each ion.

The solvation number in a pure solvent is a very important factor in this model. The solvation number is used to calculate the liquid effective mole fraction and activity coefficient of vapor pressure depression in a mixed solvent. Thus, the less accurate is the solvation number, the less accurate is the prediction. Therefore, in order to determine the solvation number more accurately, we investigated the relation between the solvation number and correlation error, in a binary solvents + salt system.

RESULTS AND DISCUSSION

Figure 5 shows the results obtained using the solvation numbers based on the Stokes' radii. Each ion solvation number is shown, which is correlated with the numbers of Stoes' radii using the salt effect data. The average absolute error of vapor composition and boiling points are also shown.

The y-x and T–x–y curves obtained using the correlated solvation numbers from observed data are shown in Fig 5. Here it can be seen that the solvation model very effectively shows the salt effect, through the change in activity coefficients, which vary from those of the no-salt system. The activity coefficient of methanol increases, whereas that of water decreases. The raising of the boiling point is also quite well correlated.

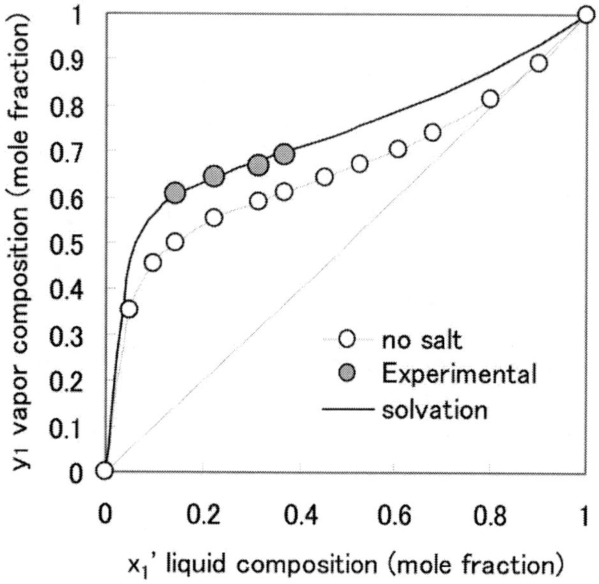

Ion salvation numbers correlated by salt effect data and reported by Marcus, and accuracy

	Correlated		Marcus	
	H_2O	EtOH	H_2O	EtOH
NH_4^+	3.3	2.7	4.6	3.7
Cl^-	2.9	1.8	3.9	3.0

Accuracy

AAD T(K)	1.4	AAD y1:mole fraction	0.006

Figure 5. The Solvation Model Applied for Ethanol-water-NH_4Cl System at 101.3 kPa

(salt concentration x_3=0.08, Jacques *et al.*, 1974)

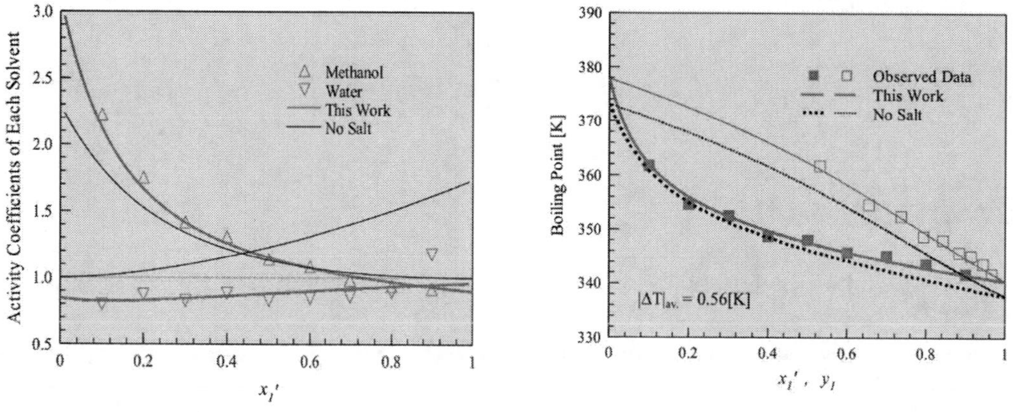

Figure 6. The Solvation Model Applied for Methanol+Water+$CaCl_2$ System at 101.3 kPa

Table 2 compares predictions from the three kinds of solvation numbers. From the table, it can be seen that results from each of the three are satisfactory. The best accuracy is given by the solvation number determined from the observed result. The result from the solvation number determined from the data on vapor pressure lowering of pure solvent + salt has medium accuracy. However, even though the accuracy is not as good, this result is quite significant, since the method can predict the salt effect. The result predicted by the solvation number determined by Stoke's radii is fairly good. The solvation number is at a fixed concentration of salt of infinite dilution. For this reason, it is difficult to obtain an accurate result; however, the model's consistency is demonstrated.

For binary solvents + salt systems, we correlated with satisfactory accuracy a total of 81 data sets for isobaric data and 16 data sets for isothermal data. For the isobaric system, the average error calculated for the vapor phase composition was 0.018 mole fraction and the average error calculated for the boiling point was 0.72 K. For the isothermal system, the average error calculated for the vapor phase composition was 0.009 mole fraction and the average error calculated for total pressure was 0.14 kPa.

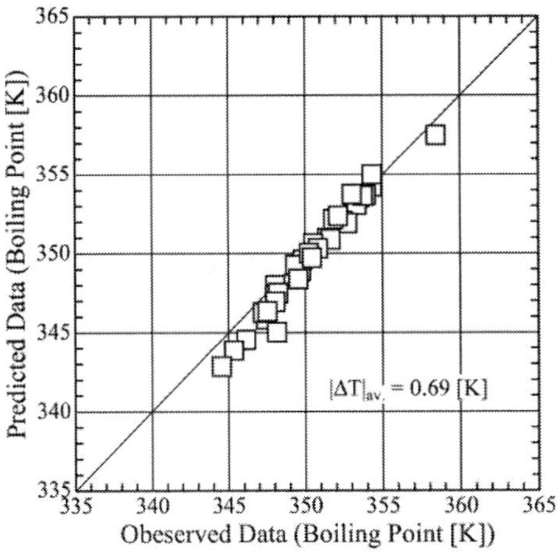

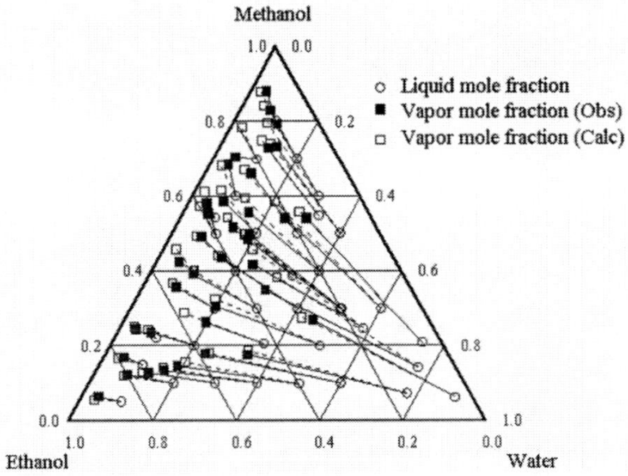

Figure 7. Prediction by Solvation Model for Methanol+Ethanol+Water+$CaCl_2$ System at 101.3 kPa

Table 2. Predicted and Correlated Result by Solvation Number

Solvent	Electrolyte,ca	salt conc.	① solvation		② stokes radii		③pure solvent	
			ΔT	ΔY	ΔT	ΔY	ΔT	ΔY
Methanol(1)-Water(2)		at 101.3kPa						
	KCl	m=0.5	0.489	0.026	0.680	0.032	3.081	0.080
		m=1.0	0.895	0.030	1.090	0.036	2.083	0.067
		m=2.0	0.715	0.020	1.990	0.058	0.721	0.025
	LiCl	m=1.0	0.945	0.017	1.146	0.018	0.648	0.038
		m=2.0	0.530	0.013	0.999	0.023	1.067	0.023
		m=4.0	0.673	0.022	6.022	0.060	0.559	0.022
	NaF	m=0.25	0.687	0.022	0.681	0.025	2.534	0.071
		m=0.5	0.594	0.014	0.735	0.019	2.870	0.076
		m=1.0	0.498	0.010	0.765	0.018	0.969	0.030
	$NaNO_3$	5mol%	0.986	0.012	1.834	0.022	1.834	0.059
		7mol%	0.560	0.018	1.222	0.037	0.331	0.017
		8mol%	0.633	0.014	2.908	0.021	0.498	0.015
	$CaCl_2$	2mol%	0.739	0.010	0.848	0.014	0.781	0.075
		4mol%	0.658	0.006	0.856	0.031	0.762	0.038
		10mol%	1.455	0.010			1.530	0.022

Solvent	Electrolyte,ca	salt conc.	① solvation		② stokes radii		③pure solvent	
			ΔT	ΔY	ΔT	ΔY	ΔT	ΔY
Ethanol(1)-Water(2)		at 101.3kPa						
	NH_4Cl	3mol%	1.294	0.005	0.917	0.021	0.915	0.019
		6mol%	0.553	0.008	0.955	0.009	0.423	0.019
		8mol%	0.940	0.013	4.226	0.049	1.551	0.023
	$CaCl_2$	16.7wt.%	1.484	0.026	1.427	0.069	1.766	0.077

Solvent	Electrolyte,ca	salt conc.	① solvation		② stokes radii		③pure solvent	
			ΔT	ΔY	ΔT	ΔY	ΔT	ΔY
Acetone(1)-Methanol(2)		at 101.3kPa						
	LiCl	0.5mol%	0.264	0.013	0.285	0.017	1.008	0.134
		0.75mol%	0.234	0.011	0.296	0.016	0.915	0.121
		1mol%	0.211	0.012	0.255	0.019	0.906	0.115
		3mol%	0.349	0.016	0.446	0.029	0.542	0.043
	NaI	2mol%	0.371	0.011	0.159	0.013	0.382	0.031
		3mol%	0.433	0.014	0.238	0.018	0.367	0.035
		4mol%	0.535	0.017	0.334	0.021	0.405	0.038
		5mol%	0.870	0.027	0.829	0.028	0.890	0.066
	$CaCl_2$	0.5mol%	0.348	0.011	0.250	0.043	0.887	0.096
		1mol%	0.647	0.039	1.144	0.067	0.647	0.039
		2mol%	0.194	0.030			0.194	0.030

Solvent	Electrolyte,ca	salt conc.	① solvation		② stokes radii		③pure solvent	
			ΔT	ΔY	ΔT	ΔY	ΔT	ΔY
2-Propanol(1)-Water(2)		at 101.3kPa						
	$CaCl_2$	m=0.4	0.517	0.018			0.593	0.021
		m=0.8	0.961	0.017			2.071	0.105
		m=1.2	1.807	0.012			3.224	0.009

average	① solvation		② stokes radii		③pure solvent	
	ΔT	ΔY	ΔT	ΔY	ΔT	ΔY
	0.699	0.017	1.198	0.030	1.150	0.051

Next we examined the applicability of the modified model to prediction for a multi-solvents system with a salt. Using the solvation number in a pure solvent + salt system determined from a binary solvent + salt system, we estimated the salt effect of the methanol + ethanol + water + $CaCl_2$ (16.7 wt.%) system at 101.3 kPa for 38 data sets.

The systems used in determining the solvation number of each solvent in a ternary system are for methanol and water, and for ethanol. Fig. 7 shows the result predicted for a bubble point of a system with an average absolute deviation of 0.85 K. This figure also shows the result predicted for the vapor composition of the system in mole fraction, with an average absolute deviation of 0.021 for methanol and 0.020 for ethanol.

CONCLUSION

Our method has the following advantages:

1. Our method can use the solvation number, which is independently obtained from Stokes' radii.
2. The solvation number can be obtained from the vapor pressure lowering data of a pure solvent with salt, or from the salt effect data.
3. We can predict VLE from the solvation numbers.

Moreover, our model expresses the salt effect more meaningfully. It can explain and demonstrate thermodynamically the mechanism of the salt effect on VLE. A further notable advantage of our model is that the visual interpretation of our parameters reflects more realistically the actual chemical structure of the liquid solution, making it easier to comprehend. We have applied our method for almost all systems reported in the literature with satisfactory results.

Using the Activity Coefficient Model (ACM) for Electrolyte Solutions has a drawback – namely, that it was originally designed for non-electrolytes. Thus, to modify it for electrolyte solutions is a mismatch. Our solvation model has no such problem. Most importantly, the ACM can only correlate or fit data; it cannot predict. Our model can predict the salt effect. And, it can do this using only the solvation numbers from a pure solvent and a salt.

LIST OF SYMBOLS

$a_{solvent}$ activity for lowering of vapor pressure of pure solvent system
n_{salt} mole number of salt
$n_{solvent}$ mole number of solvent
P total pressure of system (kPa)
$P_{solvent}$ total pressure of system as solvent system (kPa)
P_I vapor pressure of pure component i (kPa)
S_{i0} solvation number between solvent and salt
S_0 solvation number of salt with pure component of mixed solvent system
x mole fraction of ith solvent existing salt
x_i' mole fraction of ith solvent at salt free basis
x_{ia}' effective mole fraction of ith solvent at salt free basis
x_{salt} mole fraction of salt
$x_{solvent}$ mole fraction of solvent

GREEK SYMBOLS

γ_I activity coefficient of conventionally defined
γ'_I activity coefficient for effective mole fraction of i-th solvent
$\gamma_{i,solvent}$ activity coefficient for lowering of vapor pressure of i-th solvent
$\gamma_{mix,solvent}$ activity coefficient for lowering of vapor pressure of mixed solvent system

SUBSCRIPT

i i-th solvent

REFERENCES

Achard, C., C.G. Dussap, J.B. Gros, Fluid Phase Equilib. 98 (1994) 71–89.

Boone J.E., R.W. Rousseau, E.M. Schoenborn, Adv. Chem. Ser. 155 (1976) 36–52.

Chen, C.-C., H.I. Britt, J.F. Boston, L.B. Evans, AIChE J. 28 (1982) 588–596.

Chen, C.-C., L.B. Evans, AIChE J. 32 (1986) 444–453.

Ciparis, J.N., Data of Salt Effect in Vapour–Liquid Equilibrium, Lithuanian Agricultural Academy, Kaunus, 1966.

Gmehling J., U. Onken, W. Arlt, Vapor–Liquid Equilibrium Data Collection (Part 1a), Vol. 1, DECHEMA, Frankfurt, 1981.

Gmehling J., U. Onken, W. Arlt, Vapor–Liquid Equilibrium Data Collection (Part 2c), Vol. 1, DECHEMA, Frankfurt, 1982.

Hongo M., T. Hibino, Kagaku Kogaku Ronbunshu 15 (1989) 863–867.

Hongo M., M. Kusunoki, Y. Kumagae, Y. Arai, J. Chem. Thermodyn. 24 (1992) 649–651.

Iliuta M.C., F.C. Thyrion, O.M. Landauer, Fluid Phase Equilib. 130 (1997) 253–269.

Iliuta M.C., F.C. Thyrion, Fluid Phase Equilib. 103 (1995) 257–284.

Iliuta M.C., I. Iliuta, O.M. Landauer, F.C. Thyrion, Fluid Phase Equilib. 149 (1998) 163–176.

Jacques D., W.F. Furter, Ind. Eng. Chem. Fundam. 13 (1974) 238.

Kato M., T. Sato, M. Hirata, J. Chem. Eng. Jpn. 4 (1971) 308–311.

Kikic, I., M. Fermeglia, P. Rasmussen, Chem. Eng. Sci. 46 (1991) 2775–2780.

Kumagae Y., K. Mishima, M. Hongo, M. Kusunoki, Y. Arai, Can. J. Chem. Eng. 70 (1992) 1180–1185.

Kupriyaonova Z.N., V.F. Belugin, G.B. Shakhova, Zh. Prikl. Khim. 46 (1973) 234–237.

Liu W., J. Wang, K. Zhuo, C. Wang, J. Lu, Acta Chem. Sinica 56 (1998) 21–31.

Macedo E.A., P. Skovborg, P. Rasmussen, Chem. Eng. Sci. 45 (1990) 875–882.

Marcus, Y., Ion solvation, John Wiley & Sons Limited (1975)

Marcus, Y., Electrostriction, Ion Solvation, and Solvent Release onion Pairing., J.Phys.Chem.B, 109, 18541-18549 (2005)

Mock B., L. B. Evans, C. -C. Chen, AIChE J. 32 (1986) 1655–1664.

Nishi Y., J. Chem. Eng. Jpn. 8 (1975) 187–191.

Natarajan T.S., D. Srinivassan, J. Chem. Eng. Data 25 (1980) 218–221.

Ohe S., K. Yokoyama, S. Nakamura, Kagaku Kogaku 34 (1970) 325–328.

Ohe S., K. Yokoyama, S. Nakamura, Kogyo Kagaku Zasshi 72 (1969) 313–316.

Ohe S., K. Yokoyama, S. Nakamura, J. Chem. Eng. Jpn. 2 (1969) 1–4.

Ohe S., K. Yokoyama, S. Nakamura, J. Chem. Eng. Data 16 (1971) 70–72.

Ohe S., K. Yokoyama, S. Nakamura, Kagaku Kogaku 35 (1971) 104–107.

Ohe, S., Adv. Chem. Ser. 155 (1976) 53–74.

Ohe, S., Vapor–Liquid Equilibrium Data—Salt Effect, Elsevier, New York, 1991.

Ohe, S., Fluid Phase Equilib. 144 (1998) 119–129.

Ohe, S., H. Takamatsu, Kagaku Kogaku Ronbunshu 26 (2000) 275–279.

Ohe, S., Dr. Eng. Thesis, Tokyo Metropolitian University, Japan, 1971.

Ohe, S., Vapor–Liquid Equilibrium Data, Elsevier, New York, 1989.

Polka H., J. Gmehling, J. Chem. Eng. Data 39 (1994) 621–624.

Polka, J. H., Li, J. Gmehling, Fluid Phase Equilib. 94 (1994) 89–114.

Polka, H., J. Li, J. Gmehling, Fluid Phase Equilib. 94 (1994) 115–127.

Rajendran M., S. Renganarayanan, D. Srinivasan, Fluid Phase Equilib. 44 (1988) 53–73.

Rajendran M., S. Renganarayanan, D. Srinivasan, Fluid Phase Equilib. 70 (1991) 65–106.

Sander B., A. Fredenslund, P. Rasmussen, Chem. Eng. Sci. 41 (1986) 1171–1183.

Tan, T.C. Chem. Eng. Res. Des. 65 (1987) 355–366.

Tatsievkaya G.I., T.A. Vitman, T.M. Serafimov, Russ. J. Phys. Chem. 56 (1982) 1668–1670.

Yan, W., M. Topphoff, C. Rose, J. Gmehling, Fluid Phase Equilib. 162 (1999) 97–113.

Yan W., M. Topphoff, M. Zhu, J. Gmehling, J. Chem. Eng. Data 44 (1999) 314–318.

Zemp R.J., A.Z. Frabcesconl, J. Chem. Eng. Data 37 (1992) 313–316.

Structural, Energetic and Thermodynamic Studies of Acrylic (PAA, PMA and PMMA) and Allylamine (PAH) Polymers for Self Assembly

Phani G. K. Dathara and Daniela S. Mainardi. Louisiana Tech University, 911 Hergot St., Ruston, LA 71272

Organic polymer electrolytes are the most studied biocompatible materials used in applications related to implants, bio-sensors, drug delivery and tissue engineering. The goal of this research work is to study models of acrylic (PAA, PMA and PMMA) and allylamine (PAH) polymers, their interactions with supports and water solvent for their applications as biocompatible materials. Quantum Mechanical and Molecular Dynamics calculations are used to obtain ground state geometries of monomers and larger oligomers, enthalpies of formation, enthalpies of ionization, and ionization potentials. Hence, from the obtained ground state geometries of the repeat units of PAA, PMA PMMA and PAH, longer chains of respective polymers are constructed and the potential energies as well as charge distribution on the atoms calculated. Polymers in their neutral and ionized forms are modeled and the electrostatic potential derived charges on atoms are determined to have an estimate of the charge to be applied to the substrate to facilitate self-assembly. In addition to gas phase calculations, effects of solvent (water) on the polymers are also studied using Polarizable Continuum Model (PCM) calculations. Molecular Dynamics simulations are used to investigate the water sorption dynamics of the polymer chains and the electrostatic self-assembly of these polymers on titanium substrates. The percentage of water uptake by the polymers as a function of time is calculated and presented. Concentration profiles of the polymer and water in the simulated box is shown at various time steps to determine the amount of water associated around the polymer. PAA has shown higher water uptake among three polymers followed by PMA and PAH. The mean square displacement of water molecules in polymer systems of PAA, PMA and PAH are calculated and water diffusion coefficients found to be 2.3x10-3, 1.9 x10-3, and 1.3x10-4cm2/sec respectively. The electrostatic self assembly of PAA on titanium substrate is simulated using Molecular Dynamics. The model includes substrate (charge fraction=1), PAA (charge fraction=1) and water molecules (SPC model). The thickness of the layers of PAA deposited on the substrate is calculated as 1.16 ± 0.28 nm which is in agreement with the experimental reported value of 1-2nm at room temperature.

Carbon Nanotube Polymer Nanocomposites for Electromechanical System Applications

Arnab Chakrabarty and Tahir Cagin. Chemical Engineering, Texas A & M University, College Station, TX 77840

Polymer nanocomposites refer to a broad range of composite materials. Due to unprecedented improvement observed in properties of nanocomposites, research interest in this area is growing exponentially. Nanotube is one of the well known nanofiller in designing nanocomposites till date. In designing better nano-composites with tailored properties built from nanotube it is of utter importance to understand the intrinsic property of the nanotube itself. While the superiority of nanotube properties has been established, scattered results mainly in thermo and mechanical properties is the major reason of the still ongoing effort of finding out the property values both in experiments and theory. These inconsistencies keep and give room for noise in works based upon these findings. In our present work we have revisited and calculated some of these properties and tried to justify the agreement and disagreement with the available literature values. The consistency of the force field and definitions used in these calculations gives one a good idea of different property values of carbon nanotube when measured in the same scale and is more reliable to use when used with one another. The simulation results we report here looks into the properties of the pristine carbon nanotube and effect of different parameters on the same, which will later help us to correlate component properties with composite properties. In order to find these properties we have done ab-initio, molecular mechanics and molecular dynamics calculations using LAMMPS, Cerius2.0, VASP and Gaussian.

Densities and Phase Loops for Four Natural Gas-like Mixtures

Kenneth R. Hall[1], Santiago Aparicio-Martinez[2], Mert Atilhan[2], Diego Christancho[2], and Saquib Ejaz[2]. (1) Artie McFerrin Department of Chemical Engineering, Texas A&M University, MS 3122, College Station, TX 77843-3122, (2) Chemical Engineering, Texas A&M University, MS 3122, College Station, TX 77843-3122

Abstract

We have measured the densities and phase lops for four mixtures that contain methane through n-pentane, carbon dioxide and nitrogen using a magnetic suspension densimeter and an isochoric apparatus. The temperatures range from 250-450 K (-10 to 350 °F), and the pressures range up to 150 MPa (22,500 psia). We have compared these data to predictions from the equations of state normally used in the natural gas industry.

The results for the phase loops are disturbing to say the least. The equations of state miss the maximum pressure by as much as 2.5-3.0 MPa (~400 psia). This would have a significant impact upon compressor specifications. The densities deviate from AGA 8 by 1.0 % in Region 1, 0.75 % in Region 2, and 0.5 % in Region 3. Most of the data lie outside the stated accuracy of AGA 8 in regions 1 and 2, while all the data fall within the stated accuracy in Region 3.

Based upon these results, it appears that equation of state development is necessary. To accomplish this work with a reasonable hope of obtaining a truly predictive equation would require an extensive data collection project.

New Technologies for Thermophysical Properties for on-Demand Product Development: the NIST Thermo Data Engine (TDE)

Robert D. Chirico, Vladimir Diky, Chris Muzny, Eric Lemmon, and Michael Frenkel. Physical and Chemical Properties Division, National Institute of Standards and Technology, 325 Broadway, Boulder, CO 80305-3328

Rapid developments in the laboratory synthesis of numerous specialty chemicals, drug and biofuel development, as well as the increasing cost of energy and environmental compliance, has created new opportunities and challenges for chemical process design. Amongst these, one of the most difficult and principal challenges is to create software capabilities to support efficient and robust chemical process design for production of customized chemical structures; i.e., to create new technologies for product on-demand design. A key impediment to the development of such technologies has been reliance on long-established 'static' protocols for generating critically evaluated thermophysical and thermochemical property data that 'feed' the process simulators. As an alternative to the static approach, the new concept of dynamic data evaluation was developed at NIST. This concept requires the development of large electronic databases capable of storing essentially all experimental data known to date with detailed descriptions of relevant metadata and uncertainties. The combination of these electronic databases with expert-system software, designed to automatically generate recommended data based on the available experimental data and a network of prediction methods, allows production of critically evaluated data 'dynamically' or 'to order'. This concept contrasts sharply with static critical data evaluation, which must be initiated far in advance of need. The critically evaluated data produced by the deployment of the dynamic evaluation concept are rigorously characterized with quality assessments, thus providing the ability to propagate reliable data-quality limits to all aspects of chemical process design. ThermoData Engine (TDE) is the first software implementation of dynamic data evaluation. The software incorporates all stages of the concept implementation, including data retrieval, grouping, normalization, sorting, consistency enforcement, fitting, and prediction. TDE emphasizes enforcement of property consistency, provides for flexibility in selection of default data models depending on the particular data scenario, and incorporates a large variety of models for secondary fitting. The software will be described with particular emphasis on aspects essential for product-on-demand development. Opportunities and future plans related to this new technology will be discussed.

Web Page: www.trc.nist.gov/tde.html

THERMODYNAMIC PREDICTION OF THE CHAIN PROPAGATION PROBABILITY (α) FOR THE FISCHER-TROPSCH REACTION

Abstract

The Fischer-Tropsch synthesis is a surface polymerization reaction. The product distribution of hydrocarbons has been described by the Anderson-Schulz-Flory (ASF) distribution. The ASF model describes the entire product range by a single parameter, α, the probability of the addition of a carbon intermediate to a chain. However, significant deviations are reported in literature. Several models have been proposed to explain the observed two-alpha product distribution but most studies suggest that the reversible olefin termination path is responsible for the two-alpha product distribution. However, most studies ignore the presence of Vapour-Liquid Equilibrium (VLE).

This work shows that by considering the effects of VLE the observed deviations from the ASF distribution can be explained. This shows that Fischer-Tropsch reaction can be modelled by a surface polymerization reaction following a single alpha ASF distribution in the liquid phase. The effect of Vapour-Liquid Equilibrium can explain the observed two-alpha product distribution for the total hydrocarbon products by assuming single alpha ASF distribution in the liquid phase.

Keywords: Fischer-Tropsch Reaction, Thermodynamic Equilibrium, Liquid Phase Modelling, Product Distribution Model, Chain propagation probability (α), ASF.

Introduction

Fischer-Tropsch (FT) process is a process in which synthesis gas, a mixture of predominantly CO and H_2, obtained from coal, biomass or natural gas is converted to a multi-component mixture of hydrocarbons. The FT process is mainly for the production of fuels although other valuable products besides fuels can be tailor-made with the FT in combination with upgrade processes. The FT synthesis is a surface polymerization reaction [1]. The accepted model is that reactants, CO and H_2, adsorb and dissociate at the surface of the catalyst and react to form chain initiator (CH_3), methylene (CH_2) monomer and water. The hydrocarbons are formed by CH_2 insertion into metal-alkyl bonds and subsequent dehydrogenation or hydrogenation to an α -olefin or paraffin, respectively.

According to Anderson [2], the product distribution of hydrocarbons can be described by the Anderson-Schulz-Flory (ASF) equation:

$$m_n = (1-\alpha)\,\alpha^{n-1}$$

with m_n the mole fraction of a hydrocarbon with chain length n and the growth probability factor α is independent of n. α determines the total carbon number distribution of the FT products. Hence, a plot of the logarithm of the molar concentration versus the carbon number would produce a straight line plot whose slope is related to alpha (α), the chain propagation probability. The total product yield decreases exponentially with increasing chain length.

Most studies assume the Anderson-Schulz-Flory (ASF) distribution to model the product selectivity. The ASF model describes the entire product range by a single parameter, α, the probability of the addition of a carbon intermediate to a chain. However, significant deviations are reported in literature [3, 4]. Several models have been proposed to explain the observed two-alpha product distribution but most studies suggest that the reversible olefin termination path is responsible for the two-alpha product distribution [5].

According to Caldwell and Van Vuuren [6], liquid easily forms inside the FT reactor for high alpha but a lot of researchers have ignored the presence of Vapour-Liquid Equilibrium (VLE) in their Reactor Models. This study considers VLE in the development of a model that explains observed deviation from the classical ASF model.

Reactor modelling

Considering a Reactor with in which CO and H_2 continuously flow in at constant rates in a ratio of 1:2. Vapour phase products, i.e., unconverted CO and H_2, water, and volatile hydrocarbons, continuously flow out. Liquid phase products are either removed continuously or allowed to accumulate in the reactor for some time interval and drained periodically, as necessary, so that the reactor is operated at a nearly constant liquid level. The drained liquid is added to the vapour phase product to obtain the total reaction product for the sampling period.

Assuming that CO and H_2 react stoichiometrically to produce hydrocarbons (exclusively paraffins for simplicity) and water following a single α ASF distribution. The Fischer-Tropsch reaction can be written as:

$$nCO + (2n+1)H_2 \rightarrow C_nH_{2n+2} + nH_2O$$

Assuming that the vapour and the liquid products are in thermodynamic equilibrium, their compositions are related by:

$$y_i = k_i x_i$$

Also assuming that the Vapour-Liquid Equilibrium can be described by Raoult's Law gives:

$$y_i P = x_i P_{vap,i}(T)$$

Therefore the product composition could be calculated by using the constraints:

$$\sum x_i = 1 \quad \text{and} \quad \sum y_i = 1$$

With knowledge of the initial feed flow rates, the liquid composition, vapour composition, vapor flow rate, and the amount of liquid in the reactor can be calculated at any time-on-stream.

The simulation was performed using a computer program for the model described. For simplicity of simulation, it was assumed that hydrocarbon products are linear paraffins from C1 to C30.

Discussions and Results

Case 1

Considering that the total (vapour and liquid) hydrocarbon products are described by a single α ASF distribution. The case presented here is for α = 0.85, the reaction temperature of 220 °C and the pressure of 20 bar. Hence the model described above was used to calculate the resulting vapour flow rates and compositions. The results obtained are shown in Figure 1 below.

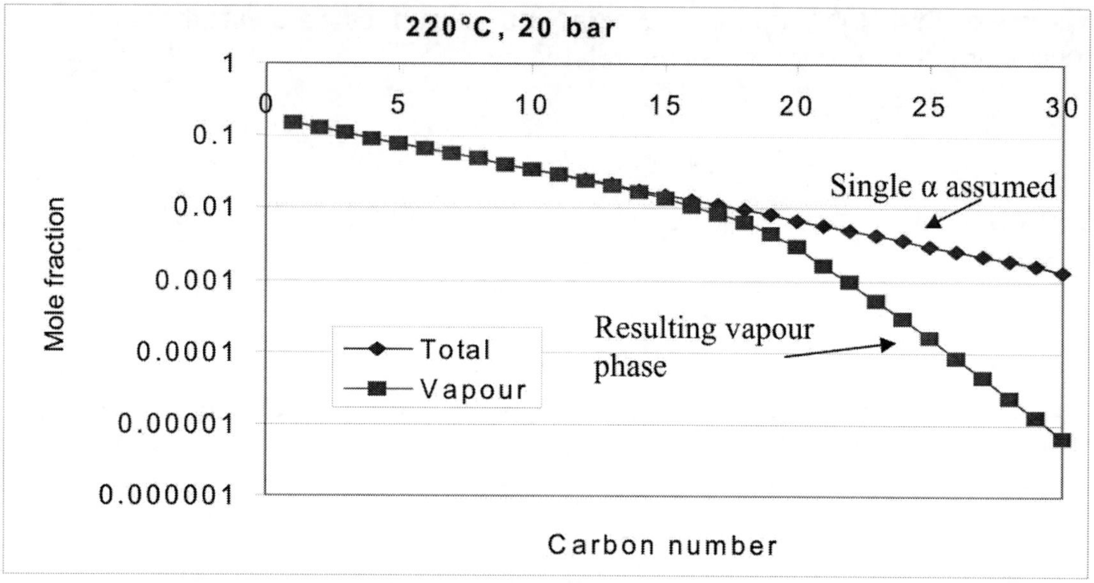

Figure 1: Observed deviation in the vapour phase for assumed single alpha ASF distribution for the total hydrocarbon products.

The vapour phase modelling always shows a negative deviation from the ASF distribution. This does not explain the observed two-alpha plot with a positive deviation.

Case 2

Considering that the vapour phase is described by a single α ASF distribution and calculating the resulting total product distribution shows the deviation presented in Figure 2.

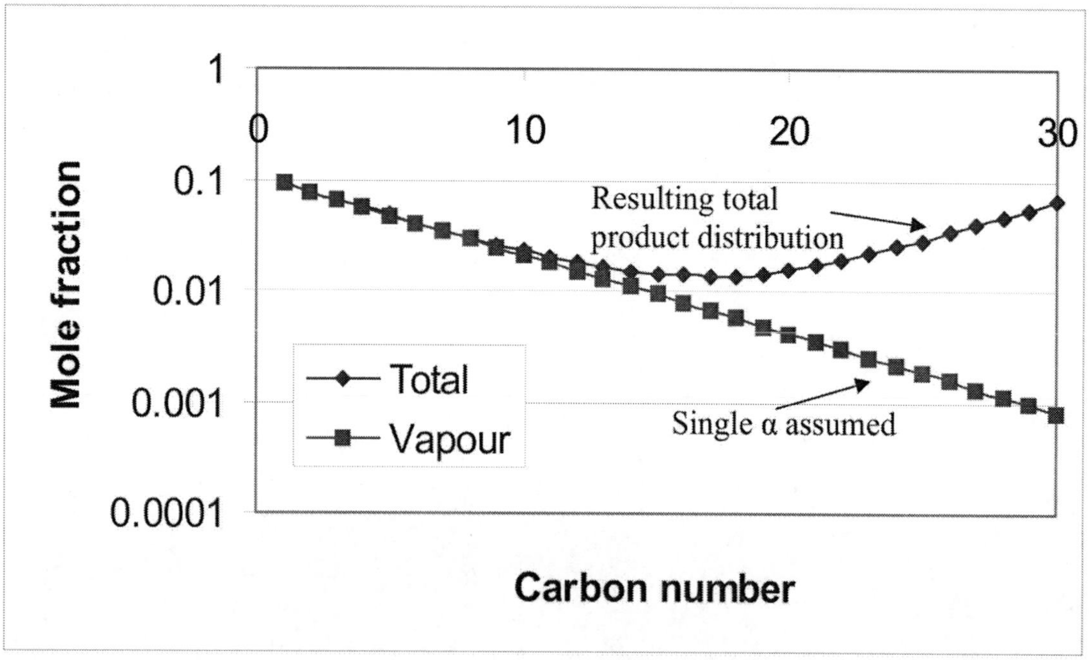

Figure 2: Observed deviation in the total product distribution for assumed single alpha ASF distribution in the vapour phase.

The observed deviation does show a positive deviation from the classical single alpha ASF distribution. However, the concentration of higher carbon numbers keeps increasing indefinitely. This also does not explain the observed deviation.

Most researchers have only considered these two cases. But once liquid is formed in the reactor the Fischer-Tropsch product would be in the liquid phase. Hence if Fischer-Tropsch product follows the ASF distribution, it should be in the Liquid phase.

Case 3

Therefore considering single alpha ASF in the Liquid phase produces the results presented in Figure 3.

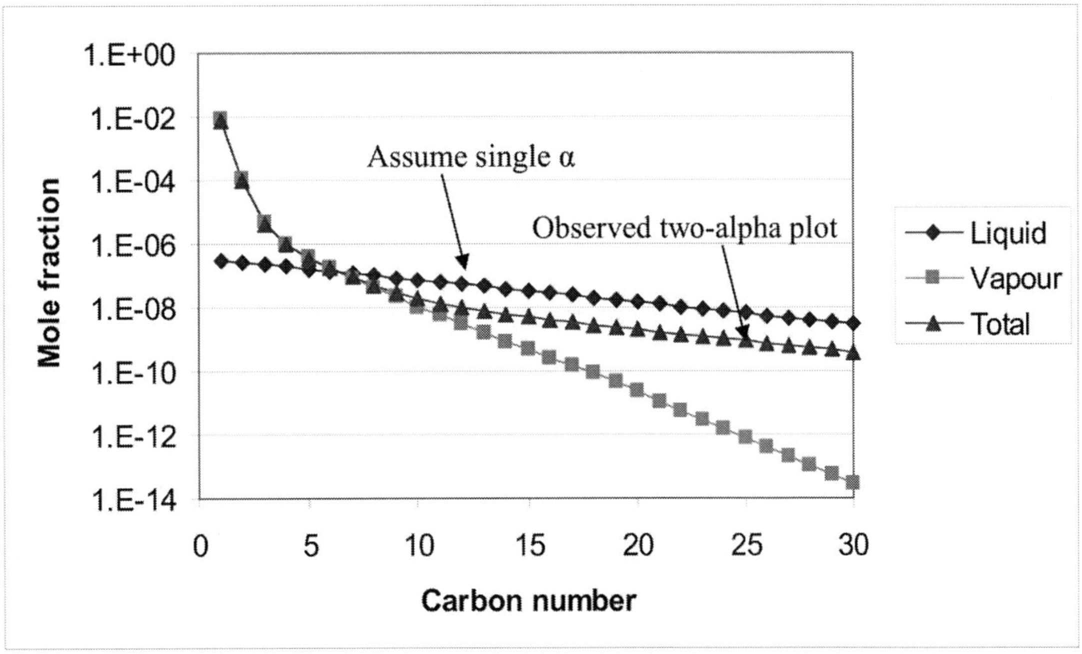

Figure 3: Observed deviation in the total product distribution for assumed single alpha ASF distribution in the liquid phase.

This clearly shows the observed deviation from the single alpha ASF distribution. Hence, considering single alpha in the liquid phase does explain the two-alpha product distribution.

Conclusion

The Fischer-Tropsch reaction is a surface polymerization reaction following a single alpha ASF distribution in the liquid phase.

The effect of Vapour-Liquid Equilibrium fully explains the observed two-alpha product distribution for the total hydrocarbon products by assuming single alpha ASF distribution in the liquid phase.

References

[1] Anderson R. B.; Friedel H. and Scorch H. H., Fischer-Tropsch reaction mechanism involving stepwise growth of carbon chain, *J. Chem. Phys.*, **1951**, *19*, 313-319.

[2] Anderson R. B., *Catalysts for the Fischer-Tropsch Synthesis*, Reinhold New York, **1956**, *4*, 337.

[3] Shi B. and Davis B. H., Fischer-Tropsch synthesis: accounting for chain length related phenomena, *Appl. Catal. A*, **2004**, 61-69.

[4] Donnelly T. J.; Yates I. C. and Satterfield C. N., Analysis and Prediction of Product Distributions of the Fischer-Tropsch Synthesis, *Energy Fuels*, **1988**, *2*, 734-739.

[5] Van der Laan G. P. and Beenackers A. A. C. M., Kinetics and Selectivity of the Fischer-Tropsch synthesis: A literature review, *Catal. Rev. -Sci. Eng.*, **1999**, *41*, 255

[6] Caldwell L. and Van Vuuren D. S., On the formation and composition of the liquid phase in Fischer-Tropsch reactor, *Chem. Eng. Sci.*, **1986**, *41*, 89-96.

Classical Molecular Dynamics on Composite Systems of Clathrate-Methane-Water-Kinetic Inhibitor

DIEGO A. GOMEZ GUALDRON and Perla Balbuena. CHEMICAL ENGINEERING, TEXAS A&M UNIVERSITY, Calle 7 #19-47, BUCARAMANGA, Colombia

Classical molecular dynamics simulations are performed for chlathrate-methane-water-inhibitor systems using different inhibitor/water ratios and a 51262cavity with its water constituents 'frozen' at their initial positions. Vinylpyrrolidone (VP), Vinylvalerolactam (Vval), L-Proline (LP), 1-Formylpyrrolidine (1FP), Dimethylaminoethyl methacrylate (DMAEMA) are used as inhibitors. The goal is to get some insights on the behavior of the systems in the early stages of chlathrate formation where the inhibitor is thought to perform its main inhibition effect. The Dreiding Force Field is used for methane and the inhibitor and SPC/E for both free water and water constituting a chlathrate. Before performing the simulations for the chlathrate systems, evidence was searched to find if the combination of force field used reproduced certain phenomena such as the cage effect for the methane-water system, and the agglomeration of water in chlathrate-like structures at the concentrations used with the 'frozen' cavity. The decision of fixing the crystal was based upon Hartree-Fock simulations where the inhibitor was found to adsorb on the cavity without altering it. The systems are analyzed via visualization of the trajectories, examination of the radial distribution functions and of the mean square displacements. In most cases the inhibitors are found to adsorb on the cavity via hydrogen bonds between pendant hydrogen atoms of the cavity and the carbonyl group of the inhibitor. While adsorbed on the crystal, the inhibitor also forms hydrogen bond with the initially free water showing evidence of the possible action of inhibitor via disrupting the chlathrate network and becoming part of the network itself. No waters are found to locate around the hydrophobic moiety, suggesting the hypothesis of a delay in the growth onset caused by the presence of the adsorbed inhibitor covering faces of the incipient crystals which are in this way refrained to keep nucleating by contact to other crystals.

ABSTRACT

A thermodynamic consistency test developed by the authors is applied to high pressure phase equilibrium data of ternary mixtures containing a compressed gas and two solid solutes. The experimental data have been modeled using the Peng-Robinson equation of state with classical van der Waals mixing rules. These type of mixtures are ussualy highly asymmetric and present inherent difficulties of analysis, in particular because of the dispersion of the data and the very low concentration of the solutes in the compressed gas phase. Despite this, the model employed allows correlating the solubility of the solids in the high pressure gas with deviations much lower than other models. The complex mixture carbon dioxide+capsaicin+b-carotene is used as a study case. It is shown that the thermodynamic test can be applied with confidence and consistency or inconsistency of the data can be accurately determined. <>INTRODUCTION

Solubility data of solids in high pressure gases reported in the literature usually present large discrepancies, which have been attributed to different reasons by different researchers. Even though we cannot be certain on the true reasons for the large discrepancies, we need to decide whether a set of solubility data is acceptable enough as to be used in process design, analysis and simulation. Therefore, methods to test inherent inaccuracies of the experimental data have become necessary. The so-called "*thermodynamic consistency tests*", based on the Gibbs-Duhem equation, are the usual ways of checking experimental phase equilibrium data. Methods applicable to low pressure data have been widely published [1], but studies on thermodynamic consistency tests for high pressure phase equilibrium data are scarce and the only attempt for ternary mixtures we know is that recently presented by the authors [2]. The cases of interests in this study are mixtures containing two solid solutes dissolved in a high pressure solvent. With some appropriate assumptions the mole fraction of the solute "i" in the high pressure solvent, or solubility, at the temperature T, can be reduced to [3, 4]:

$$y_i = \frac{P_i^{sub}\, e^{\frac{V_i^s}{RT}(P - P_i^{sub})}}{P\varphi_i}$$

In eqn. (1), j_i is the fugacity coefficient of the solid component in the high pressure gas, P_i^{sub} is the sublimation pressure of the pure solid, V_i^s is the molar volume of the solid, T is the absolute temperature, and R is the ideal gas constant.

For a multicomponent homogeneous gas mixture at constant temperature, the Gibbs-Duhem equation can be written as [5]:

$$\left[\frac{(Z-1)}{P}\right] dP = \Sigma\; y_i d(\ln \varphi_i)$$

being y_i the mole fraction of component "i" in the gas phase, P is the system pressure, Z is the compressibility factor of the gas mixture, and j_i is the fugacity coefficient of component "i". If more than one solute is present in the high pressure gas mixture, one equation for each solute component must be written. For a given solute "j" the equation is:

$$\int \frac{1}{P y_j} dP = \sum_{i=1}^{2} \int \frac{y_i}{(Z-1)\varphi_i y_j} d(\varphi_i)$$

The Peng-Robinson equation with the van der Waals mixing rule [6] has been used in this study to correlate the experimental ternary high pressure solubility data and to determine the fugacity coefficients j_I needed in eqn. (3). The model is designated here as PR/vdW.
<>Consistency Criteria

To define the criteria for thermodynamic consistency it is first required that the model be able to correlate the data within acceptable deviations. The model is accepted and the consistency test is applied if deviations between calculated and experimental solubility values for each and every point are within -20% to 20%. After the model is found appropriate, it is required that the Gibbs-Duhem equation is fulfilled [2, 7]. In eqn. (3 the left hand side is designated by A_{Pj} and the expression inside the summation term on the right hand side by A_{jj}. That is:

$$A_{Pj} = \int \frac{1}{P y_j} dP \qquad A_{\varphi j} = \sum_{i=1}^{2} \left[A_{\varphi j}\right]_i$$

Thus, if a set of data is considered to be consistent A_{Pj} should be equal to A_{jj} for j=2 and j=3 within acceptable defined deviations. To set the margins of errors, an individual percent area deviation $[\%DA_j]_i$ between experimental and calculated values is defined as:

$$\left[\%\Delta A_j\right]_i = 100\left[\frac{A_\varphi - A_P}{A_P}\right]_i$$

The consistency method requires that the difference between the integral on the left hand side of eqn. (3) and the sum of the integrals on the right hand side is between defined margins (between -20% and +20%), to declare the data as being thermodynamically consistent. In a previous work on binary mixtures the authors have clearly explained how these limits of acceptance for the solubility and the areas, were determined [7]. The data are considered to be thermodynamically consistent (TC) if the deviations in correlating the solute solubility and all the individual deviations in the areas are within the defined limits. The data are considered to be thermodynamically inconsistent (TI) if the deviations in correlating the solute solubility are within the established limits but the individual deviations in the areas are outside the established limits, for more than 25% of the data points in the data set. If only some few data points (less than 25% of the original points), give deviations outside the accepted range the set of data is declared to be not fully consistent (NFC).

<>Results and DiscussioN

To the best of our knowledge the only set of data for the ternary mixture CO_2+capsaicin+b-carotene published in the literature is that of Skerget and Knez [8], data that is checked for consistency in this study. The range of pressure of the data is from 10 to 30 MPa and the range of solubility is from $2.4x10^{-4}$ to $4.5x10^{-4}$ in mole fraction for capsaicin and from $1x10^{-7}$ to $6x10^{-7}$ in mole fraction for β-carotene. The Table 1 shows the modeling results for the ternary mixture. This Table shows the deviations between experimental and calculated values for the solute solubilities and the interaction parameters k_{12}, k_{13}, *and* k_{23} calculated using ternary solubility data and the PR/vdW model. The model parameters were determined using genetic algorithms as described by the authors [2, 7], method that has been successfully applied in this study. In all cases the deviations for the gas solvent is lower than 1%, so numbers are not shown. Table 2 presents the results of the consistency test for the ternary mixture. Table 1: Calculated deviations and optimum interaction parameters for the ternary mixture CO_2(1)+capsaicin(2)+β-carotene(3) using the PR+VdW model.

T(K)	N	P(MPa)	k_{12}	k_{13}	k_{23}	$\%Dy_2$	$\|\%Dy_2\|$	$\%Dy_3$	$\|\%Dy_3\|$
298	10	10-28	0.0569	0.0693	0.0010	-6.63	16.90	-3.03	6.23
313	10	10-30	0.0715	0.0578	0.2957	-3.49	10.97	-1.64	9.33

Table 2: Details of the consistency test for the ternary systems CO_2(1)+capsaicin(2)+β-carotene(3)

T(K)	N	Range P(MPa)	Max. $[\%DA_2]_i$	Max. $[\%DA_3]_i$	Aver. $[\%DA_2]_i$	Aver. $[\%DA_3]_i$	1. Result
298	10	14-26	59.0	19.1	-----	-----	NFC
298	7	18-26	-14.7	19.1	-3.6	5.7	TC
313	10	21-30	29.0	24.6	-----	-----	NFC
313	8	23-30	13.1	-12.0	-0.7	-1.5	TC

The original data set containing ten points at both temperatures (298 and 313K) showed to be not fully consistent (NFC). This is so because the maximum deviations in the individual areas, designated as Max.$[\%DA_2]_i$ and Max.$[\%DA_3]_i$, are outside the acceptable interval of -20% to +20%. Once the data with high area deviations are removed, the remaining set of 7 data points at 298K and 8 data points at 313K showed to be consistent. The subindexes 2 and 3 indicate the average deviation with respect to solutes 2 and 3, respectively. Although the average percent deviation is not explicitly considered as a determining parameter for the decision of consistency or inconsistency, the results presented in Table 5 show the expected behavior. <>CONCLUSIONS

According to the results, the following conclusions can be drawn: i) the Peng-Robinson equation with the van der Waals mixing rule can be used to acceptable correlate the experimental ternary high pressure solubility data; ii) the numerical method based on genetic algorithms shows to be efficient for searching a global optimum; and iii) the consistency test allows to determine the thermodynamic consistency of the experimental data. **ACKNOWLEDGMENTS**

The authors thank the support of the National Commission for Scientific and Technological Research (CONICYT-Chile), through the research grant FONDECYT 1040285, the Direction of Research of the University of La Serena-Chile for permanent support through several research grants and the Center for Technological Information (CIT, La Serena-Chile), for computer and library support. <>REFERENCES

[1] Raal, J.D., Mühlbauer A.L., *Phase Equilibria: Measurement and Computation*, Taylor & Francis, Washington, DC (1988)

[2] Valderrama J.O., Robles P.A., Fluid Phase Equil., 242(1), 93-102 (2006)

[3] Walas S.M., *Phase Equilibria in Chemical Engineering*, Butterworth Publisher-USA (1985)

[4] Prausnitz J.M., Lichtenthaler R.N., Gomez de Azevedo E. (1999*), Molecular Thermodynamics of Fluid Phase Equilibria,* 3rd edition, Prentice Hall. Englewood Cliffs, NJ - USA (1999)

[5] Bertucco, A.; Barolo, M.; Elvassore, N., AIChE J., 43(2), 547-554 (1997)

[6] Peng D.Y., Robinson D.B., Ind. Eng. Chem. Fundam., 15(1), 59-64 (1976)

[7] Valderrama J.O., Zavaleta J., J. Supercrit. Fluids., 39(1), 20-29 (2006)

[8] Skerget M., Knez Z., J. Agric. Food Chem., 45, 2066-2069 (1997)

Density Collection Technique for Gas Mixtures Using an Automated Isochoric Apparatus and a Magnetic Suspension Densitometer

Kenneth R. Hall, Mert Atilhan, Saquib Ejaz, Diego Christancho, and Santiago Apaticio-Martinez. Artie McFerrin Department of Chemical Engineering, Texas A&M University, MS 3122, College Station, TX 77843-3122

Abstract

Accurate knowledge of the PVT behavior of gases is essential for the natural gas industry. The gas densities at reservoir and pipeline conditions are necessary for production, transportation and utilization of the resource. Density plays a key role in determining economic value of natural gas for custody transfer, and it is important to have accurate gas density data covering the broad range of pipeline conditions. It is also essential to have such data to validate or develop equations of state (EOS) that can calculate phase equilibrium properties and provide accurate density predictions. For pipeline conditions, accurate cricondentherm and cricondenbar values are necessary to avoid two-phase flow (liquid + gas). In our laboratory, we have constructed an automated isochoric apparatus to measure densities and phase envelopes, and we have commissioned a magnetic suspension densitometer (MSD) for density measurements. Using the isochoric apparatus, we collect isomolar pressure vs temperature data. The isochoric apparatus can operate at temperatures ranging from 100 K to 500 K and at pressures up to 200 MPa. The temperature gradients across the cell are less than ±0.002 K. We measure pressure with 0.01% accuracy and 0.0001% resolution. The MSD can measure isothermal densities up to 200 MPa between 200 and 500 K. Temperature stability is ±5 mK and pressures have an uncertainty of ±0.002%. The MSD can provide densities accurate to ±0.06%. By crossing the isomolar PT lines obtained from isochoric apparatus with isothermal density lines obtained from the MSD, we can determine the densities of the isomolar lines at intersections with the isothermal lines. Because the volume of the isochoric cell changes with temperature and pressure, the density of an isomolar line is not quite constant. The actual densities result from applying the volume distortion equation. It is necessary to have accurate material property coefficients to correct the volumes accurately. By using our density collection technique, we determine our isochoric cell material parameters experimentally. The procedure is to use the known isothermal densities at thee intersection points with isomolar lines to determine the coefficients in the volume distortion expression. We illustrate the technique with experimental data for a pseudo-natural gas mixture.

A Critical Comparison of Manual and Automated Equation of State Production

Chris D. Muzny, Vladimir Diky, Eric Lemmon, Robert D. Chirico, and Michael Frenkel. Physical and Chemical Properties Division, National Institute of Standards and Technology, 325 Broadway, Boulder, CO 80305-3328

Chemical process simulation and design hinges on the availability of accurate thermophysical and transport properties for all chemical species present. By far the most useful representation of thermophysical properties is an accurate equation of state. However, the production of accurate equations of state is a complex process that depends on the availability of a large set of high quality, evaluated data. The traditional path to an equation of state is for a data evaluator to produce parameters that fit a complex equation of state to a chosen data set. This method requires a highly experienced thermodynamicist and can be very time consuming. Recently, the Thermodynamics Research Center at the National Institute of Standards and Technology has introduced a software system for the evaluation of thermophysical property data called ThermoData Engine (TDE). While the initial release of TDE included only single property data correlation capabilities along with subsequent consistency checks, TDE 2.0 has the additional capability of automatically fitting all property data for a given compound to a chosen equation of state. A comparison of the performance of TDE's automated equation of state production against a set of equations of state produced at NIST using more traditional methods will be presented. The various performance metrics chosen to evaluate the quality of the fit will be discussed and the results for both equation of state production methods when fitting a specific Span-Wagner equation of state will be given. Additionally, automated equation of state production for compounds for which data are relatively sparse and no current critically evaluated equation of state is available will be presented.

A New Mixing Rule to Model the Solubility of Solids in Supercritical Fluids

L. Antonio Estévez and Zalleris O. Escobar
University of Puerto Rico
Department of Chemical Engineering
Mayagüez PR 00681-9046

ABSTRACT

Currently, supercritical-fluid technology in the pharmaceutical and microelectronics industries is increasingly applied to solve difficult processing problems. The solubility of a solute in a supercritical fluid is the most important thermophysical property that needs to be determined and modeled as a first step to develop any supercritical fluids application. This research was undertaken to develop a reliable mathematical model to compute the solubility of solids in supercritical fluids. As a result, a new combination rule is proposed along with a novel approach to obtain general correlations for its parameters. The new combination rule is a modification of the classical van der Waals mixing rules where the binary cohesive parameter a_{12} is correlated in terms of the reduced pressure. A database containing experimental solubility data for 126 isotherms was used in this study. Half of the isotherms (63) were judiciously selected to develop the correlations in the new combination rule. A correlation is proposed in which the crossed cohesive parameter a_{12} of the equation of state is expressed in terms of the reduced properties and the ratio a_2/a_1. The rest of the isotherms (63) were then used to validate the results. Detailed error calculations were carried out for different thermodynamic models that included the Peng-Robinson and Patel-Teja equations of state and van der Waals, cubic, and Rao mixing rules. The conclusion, after comparing the calculated errors for various models, was that the best results were obtained for the Patel-Teja EoS and the new mixing rule proposed here. This work is a significant contribution in the field in two ways. First, it provides a specific correlation that gives excellent values of solubility. Secondly, it proposes a novel approach that can be extended to other mixing rules and may result in a fully predictive method.

PREAMBLE

The quest of a fully predictive model for the solubility of solids in supercritical fluids is the dream of any theoretician in the field. In this paper, an effort in this direction is presented, which represents a new approach to the problem. The approach consists in modifying a mixing rule by establishing a relationship for the crossed-cohesive parameter and pressure. This approach can be applied to any mixing rule and in this work was used for the van der Waals mixing rule. Also, the proposed mixing rule can be used in conjunction with any equation of state (EoS). In this work, two EoS were selected: the Peng-Robinson EoS and the Patel-Teja EoS. Due to space limitation, the rest of this paper will focus on the detailed explanation of the proposed approach. A review of the state of the art to place this contribution in a current context is postponed to a future publication.

EXPERIMENTAL DATABASE

To assess the relative quality of the different thermodynamic models, a database comprising five solvents and twelve solutes was used in this work. The temperatures ranged from 12 [°C] to 85 [°C]. Tables 1 and 2 show these compounds along with the values of their physical properties. The supercritical solvents are listed in Table 1 while the solutes are in Table 2.

Table 1. Properties of the Supercritical Solvents

Solvent	M [g/mol]	P_c [MPa]	T_c [K]	ω
chlorotrifluoromethane	104.459	3.92	301.95	0.108
carbon dioxide	44.010	7.38	304.25	0.225
ethane	30.070	4.88	305.35	0.098
ethylene	28.054	5.03	282.40	0.089
fluoroform	70.013	4.88	299.15	0.275

Table 2. Physical Properties of the Solutes

Solute	M [g/mol]	P_c [MPa]	T_c [K]	ω	A	B [K]	$10^3 \times V^{sol}$ [m^3/mol]
benzoic Acid	122.124	4.50	752.00	0.6200	14.408	4618.1	0.0965
naphthalene	128.174	4.05	748.15	0.3020	13.583	3733.9	0.1103
2,3-dimethynaphtalene	156.23	3.22	785.00	0.4240	14.0646	4302.5	0.1547
2,6-dimethynaphtalene	156.23	3.22	777.00	0.4201	14.4286	4419.5	0.1547
1,4-naphthoquinone	158.16	4.12	792.15	0.5760	14.735	4739.4	0.1112
hexamethylbenzene	162.276	2.38	752.00	0.4980	13.1336	3855.0	0.1526
fluorene	166.23	2.99	821.00	0.4070	14.2046	4561.8	0.1393
anthracene	178.234	3.12	869.30	0.3531	12.147	4397.6	0.1426
phenanthrene	178.24	3.17	882.55	0.3299	14.631	4873.4	0.1512
acridine	179.22	3.19	883.15	0.4980	13.721	4740.1	0.1783
hexachloroethane	236.74	3.45	714.60	0.1630	10.6322	2600.94	0.1132
triphenylmethane	244.34	2.24	863.00	0.5760	14.7858	5228.0	0.2409

NEW COMBINATION RULE

Rationale

In general, mixing rules refer to the way the mixture parameters of any EoS are related to the pure-component parameters. This relationship is normally expressed by two equations: one relating the mixture parameter in terms of binary parameters (e.g., a_{mix} in terms of a_{ij} and mole fractions) and another relating the binary (or crossed) parameters to pure-component parameters. The second one is independent of composition and is normally refer to as combination rule. In this work, a new combination rule is proposed for parameter a of the EoS (any EoS) in binary systems. The approach is as follows. An "experimental" value of a_{12} is computed from the experimental solubility data. One value of a_{12} is obtained for each data point in a given isotherm. The available database described above contains 126 isotherms. To simplify this calculation, infinite dilution is assumed, i.e., y_1 = 1, y_2 = 0, and $Z = Z_{solvent}$, where 1 refers to the solvent and 2 to the solute. Then, for each isotherm, a correlation of a_{12} with reduced pressure is obtained. Finally, the parameters in that correlation are in turns correlated to reduced temperature and the a_2-to-a_1 ratio. This overall approach is explained in detail below.

For the development of this new combination rule, half of the database (i.e., 63 isotherms) was used to develop the combination rule. The other half was used to validate the correlations. A representative sample of the solute and solvent combinations were included in the 63 isotherms chosen to develop the combination rule. The values of a_{11} and a_{22} corresponding to the pure components in the mixture were calculated in terms of the critical properties. The parameter a_{12} was obtained from the experimental data as described above, with a value of zero for the binary interaction parameters k_{12}. These values were plotted with respect to the reduced pressure. For example, the system anthracene and carbon dioxide has been studied by different authors. The corresponding "experimental" values of a_{12} for the Peng-Robinson EoS are shown in Figures 1 and 2 at two temperatures. As observed, a_{12} varies (decreases for most systems) with pressure and approaches the van der Waals value ($\sqrt{a_1 a_2}$) at high pressures. These trends definitely indicate that the approach is promising. A detailed study was conducted both for the Peng-Robinson and for Patel-Teja EoS, both with the van der Waals mixing rules, modified as proposed above.

The following dimensionless parameters were defined to normalize the values of a_{12}:

$$\xi = \frac{a_{12} - a_{11}}{a_{22} - a_{11}} \tag{1}$$

$$\psi = \frac{a_{22}}{a_{11}} \tag{2}$$

These definitions allowed representing all systems in one graph. Given the shape of the curves obtained in Figures 1 and 2, a linear approach was attempted in which ξ was plotted as a function of $1/P_r$. All 63 isotherms are shown in Figure 3, which illustrate such attempt. Definitely, the trend is more or less linear for most isotherm and, thus, this approach was pursued.

Analysis for the Peng-Robinson EoS

Overview

As mentioned above, Figure 3 shows a plot of ξ versus $1/P_r$ for the 63 isotherms selected to develop the correlation for the Peng-Robinson EoS. The $1/P_r$ dependency was used looking for a linear relationship. For each system, a regression was sought between ξ and the reciprocal of the reduced pressure, as follows:

$$\xi = C_0 + C_1\left(\frac{1}{P_r}\right) \quad (3)$$

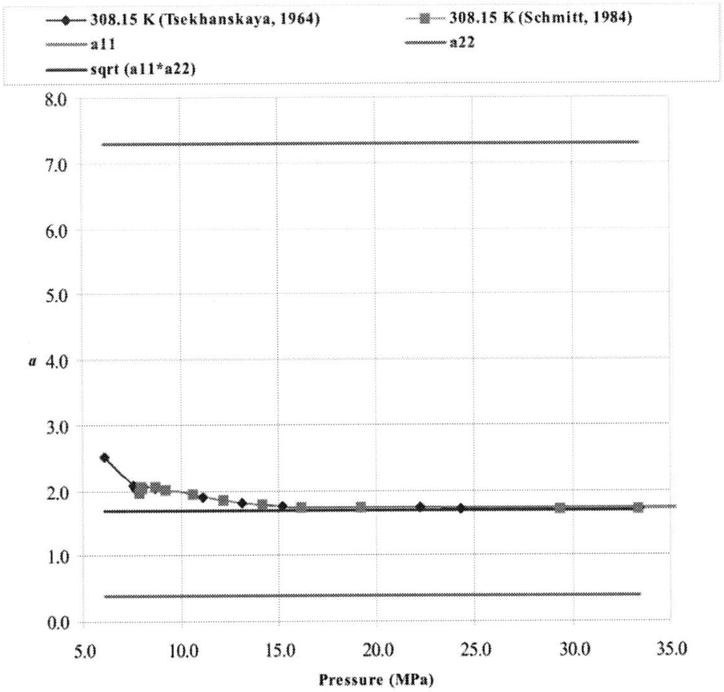

Figure 1. a_{12} **for Anthracene in Carbon Dioxide at 308.15 K**

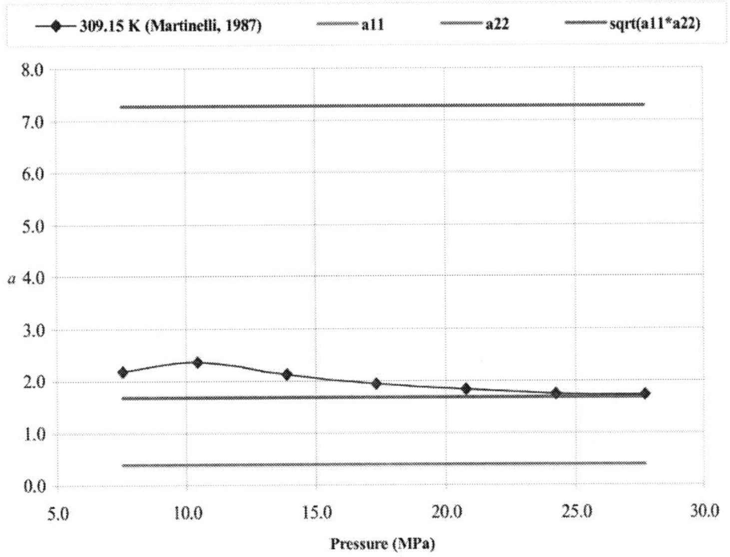

Figure 2. a_{12} **for Anthracene in Carbon Dioxide at 309.15 K**

Therefore, 63 values of C_0 and C_1 were obtained. Relationships among these variables were sought using the program Minitab. These relationships can be expressed, in general, as:

$$f_0(C_0) = F_0(\psi, T_r) \tag{4}$$

$$f_1(C_1) = F_1(\psi, T_r) \tag{5}$$

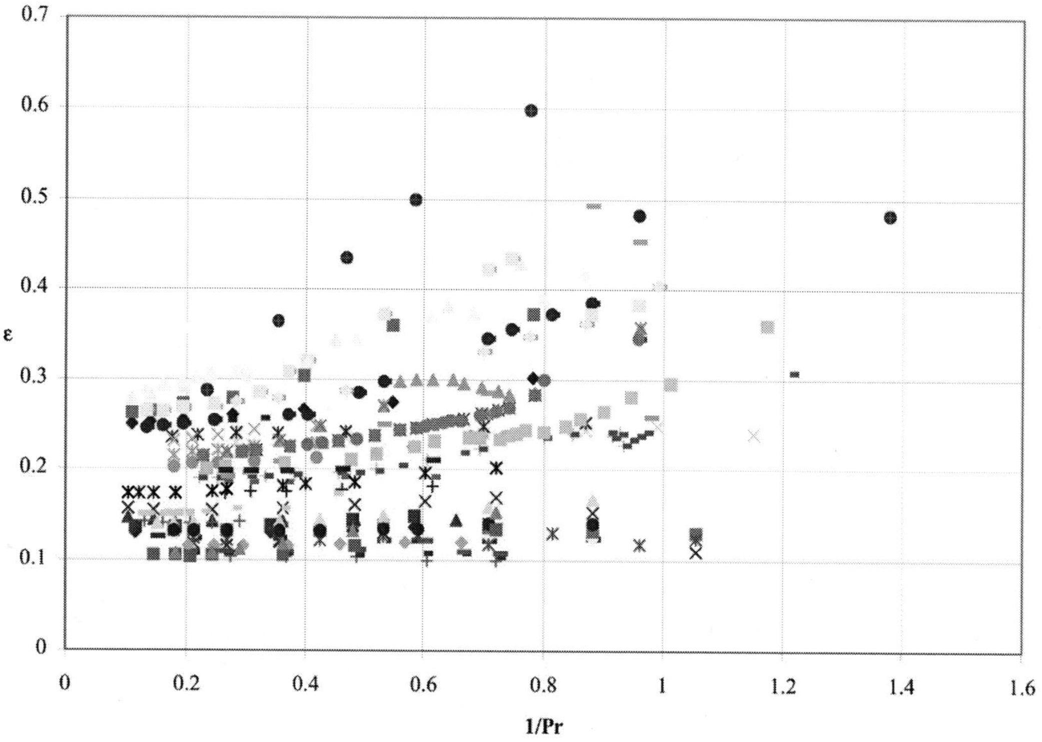

Figure 3. Values of ξ versus $1/P_r$, for the 63 Systems Studied

Correlation for C_0

As a starting point, the following function was used:

$$C_0 = p_0 + p_1\psi + p_2 T_r \tag{6}$$

The regression analysis and the corresponding statistical parameters are shown in Table 3. Additionally, Table 4 shows the analysis of variance for the variable C_0. As seen the value of R^2 is quite small and thus the correlation should be improved. The value of R^2 can be increased by making a transformation to the values of C_0, The tool Box-Cox plot, shown in the Figure 4, helped determine the most appropriate transformation. This plot provided the exponent (called lambda in Figure 4) for C_0 that gives the highest value of R^2. The suggested value is -1; this means that C_0 should be raised to this power (-1). Now, a new regression analysis was performed. Tables 5 and 6 present the results of the statistical analysis for the regression using this transformation. It can be seen that the value of R^2 increased from 48.6 % to 76.1 % and all parameters are statistically significant.

Table 3. Regression Analysis‡: C_0 Versus ψ and T_r

Predictor	Coefficients	SE Coefficient	T	p-value
p_0	0.13448	0.09073	1.48	0.144
p_1	–0.00378	0.000503	–7.52	0
p_2	0.12	0.08649	1.39	0.17

‡ The values of R^2 and R^2_{adj} are 48.6 % and 46.9 %, respectively.

Table 4. Analysis of Variance for C_0 versus ψ and T_r

Source of Variation	Degrees of Freedom	Sum of Squares	Mean Square	F_0	p-value
Regression	2	0.093654	0.046827	28.39	0
Residual Error	60	0.098952	0.001649		
Total	62	0.192605			

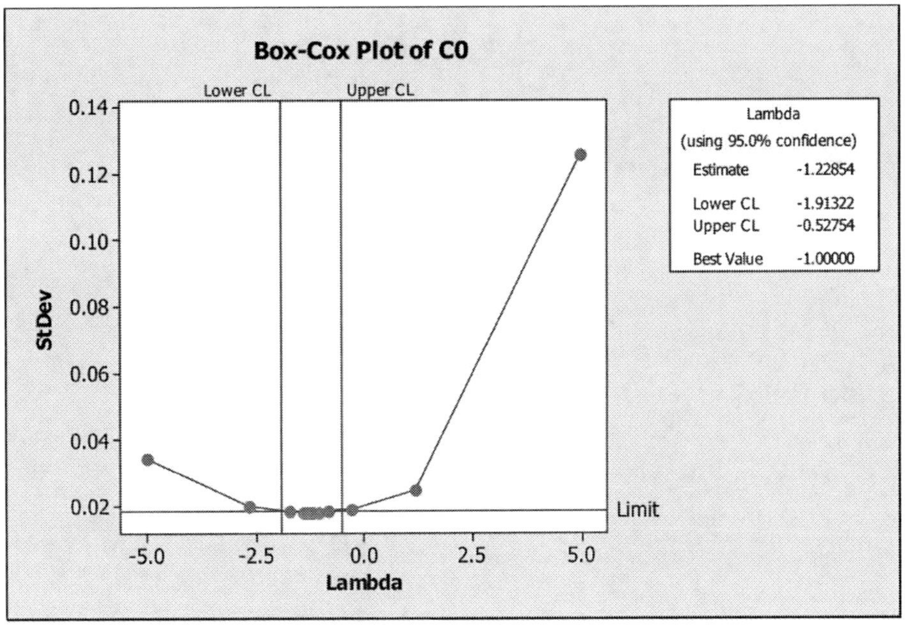

Figure 4. Box-Cox Plot for the Constant C_0

Table 5. Regression Analysis‡: Transformed C_0 versus ψ and T_r

Predictor	Coefficients	SE Coefficients	T	p-value
p_0	7.951	2.048	3.88	0
p_1	0.15648	0.01136	13.77	0
p_2	–5.013	1.952	–2.57	0.013

‡ The values of R^2 and R^2_{adj} are 76.1 % and 75.3 %, respectively.

Table 6. Analysis of Variance for Transformed C_0 versus ψ and T_r

Source of Variation	Degrees of Freedom	Sum of Squares	Mean Square	F_0	p-value
Regression	2	160.136	80.068	95.31	0
Residual Error	60	50.403	0.84		
Total	62	210.539			

Therefore, the relationship among the variables ψ, T_r, y C_0, would be expressed as follows:

$$C_0^{-1} = \frac{1}{C_0} = p_0 + p_1\psi + p_2T_r$$

or

$$C_0 = \frac{1}{p_0 + p_1\psi + p_2T}$$

Replacing the values from Table 5, the equation for C_0 becomes:

$$C_0 = \frac{1}{7.951 + 0.15648\psi - 5.013T_r} \quad (7)$$

Correlation for C_1

As a starting point, the following function was used:

$$C_1 = q_0 + q_1\psi + q_2T_r \quad (8)$$

The corresponding regression analysis is summarized in Tables 7 and 8. Again, the value of R^2 is low and, additionally, the variable T_r turned out not to be significant. Therefore, T_r was taken out from the model and a transformation to the other variable was sought to increase the value of R^2. The transformation for ψ was found by trial and error using simple transformations in Minitab and, among those tested, the best one resulted to be $1/\sqrt{\psi}$. Tables 9 and 10 present the statistical analysis for regression using with this transformation.

Table 7. Regression Analysis‡: C_1 versus ψ and T_r

Predictor	Coefficients	SE Coefficients	T	p-value
q_0	0.0727	0.1365	0.53	0.596
q_1	–0.00308	0.000757	–4.07	0
q_2	0.0548	0.1301	0.42	0.675

‡ The values of R^2 and R^2_{adj} are 22.0 % and 19.4 %, respectively.

Table 8. Analysis of Variance for C_1 versus ψ and T_r

Source of Variation	Degrees of Freedom	Sum of Square	Mean Square	F_0	P-value
Regression	2	0.063293	0.031647	8.48	0.001
Residual Error	60	0.223886	0.003731		
Total	62	0.287179			

Table 9. Regression Analysis‡: C_1 versus $1/\sqrt{\psi}$ and T_r

Predictor	Coefficients	SE Coefficients	T	p-value
q_0	– 0.11144	0.03501	– 3.18	0.002
q_1	0.7746	0.1583	4.89	0

‡ The values of R^2 and R^2_{adj} are 28.2 % and 27.0 %, respectively.

Table 10. Analysis of Variance for C_1 versus $1/\sqrt{\psi}$

Source of Variation	Degrees of Freedom	Sum of Square	Mean Square	F_0	p-value
Regression	1	0.080924	0.080924	23.93	0
Residual Error	61	0.206255	0.003381		
Total	62	0.287179			

The Box-Cox plot analysis was used to explore a transformation for C_1. However, no exponent better than 1 was found and thus no transformation for C_1 was necessary. The relationship for C_1 in terms of $1/\sqrt{\psi}$ can be thus expressed as follows:

$$C_1 = q_0 + q_1 \frac{1}{\sqrt{\psi}}$$

Replacing the values from Table 9, C_1 becomes:

$$C_1 = \frac{0.7746}{\sqrt{\psi}} - 0.1114 \tag{9}$$

Proposed mixing rule

The new combination rule for the Peng-Robinson EoS, according to the results outlined above, is given by the following equations:

$$a_{12} = \left[a_{11} + \xi\left(a_{22} - a_{11}\right)\right]\left(1 - k_{12}\right) \tag{10}$$

$$\xi = C_0 + C_1\left(\frac{1}{P_r}\right) \tag{11}$$

$$C_0 = \frac{1}{7.951 + 0.15648\psi - 5.013T_r} \tag{12}$$

$$C_1 = \frac{0.7746}{\sqrt{\psi}} - 0.1114 \tag{13}$$

where

$$\psi = \frac{a_{22}}{a_{11}} \tag{14}$$

Analysis for the Patel-Teja EoS

For the Patel-Teja EoS (PT EoS), the procedure used was analogous to the one described above. One salient difference though was that for the PT EoS, a few values of C_1 were negative; thus, some of the functions would not work properly and, more importantly, the Box-Cox plot technique would not work. To circumvent this limitation, a constant value was added to the values of C_1 before the transformation to make all values positive; this constant was 0.2. The rest of the details in the steps leading to the mixing rule have been omitted to preserve space. Only the final equations are presented here.

The new combination rule for the Patel-Teja EoS is given by the following equations:

$$a_{12} = \left[a_{11} + \xi\left(a_{22} - a_{11}\right)\right]\left(1 - k_{12}\right) \tag{15}$$

$$\xi = C_0 + C_1\left(\frac{1}{P_r}\right) \tag{16}$$

$$C_0 = \frac{2.0420}{\psi} + 0.11107T_r - 0.05248 \tag{17}$$

$$C_1 = \left(\frac{1}{1.703 - \frac{32.258}{\psi} + \frac{4.194}{T_r}} \right) - 0.2 \tag{18}$$

Where, again

$$\psi = \frac{a_{22}}{a_{11}} \tag{19}$$

RESULTS

Overview

The results presented here are mainly based on the error (AVERR) calculated according to the equation:

$$AVERR = 100 \frac{\sqrt{\frac{\sum_{i=1}^{N_p} \left(y_{i,cal} - y_{i,\exp} \right)^2}{N_p}}}{\overline{y_{\exp}}} = 100 \frac{OF}{\overline{y_{\exp}}} \tag{20}$$

For each of the models studied, AVERR was computed for every isotherm. Isotherms were separated in two groups. Group 1 contains the 63 isotherms used to obtain the correlations presented in the previous section while Group 2 contains the rest of the isotherms. Then, the values of AVERR were averaged within each of these groups. In some cases, the two groups were lumped together.

Errors for Various Models

To have an overall view of the results, Figure 5 was prepared. It shows the averaged AVERR for all 126 isotherms (i.e., the two groups described above lumped together) for various models. It is observed there that Patel-Teja with the new mixing rule (NMR) gives the best results. Moreover, a comparison of the average errors for the Peng-Robinson EoS with different mixing rules gives values between a 20.2 and 28.6%. The largest error corresponded to the Rao mixing rule (28.6%) followed by the cubic mixing rule (26.0%) and van der Waals mixing rule (20.2%). Figure 1 also shows, for the van der Waals mixing rule, a comparison of the Peng-Robinson and Patel-Teja EoS. As seen, the Patel-Teja EoS was better (16.4%) than the Peng-Robinson EoS (20.2% error).

In Figure 6, the errors of four of the models are compared by groups. It is interesting to note that in all cases the error for Group 2 are smaller than those of Group 1 when one would expect the opposite to occur. When all the isotherms are taken into account, the model Patel-Teja with the new mixing rule results in the smallest error.

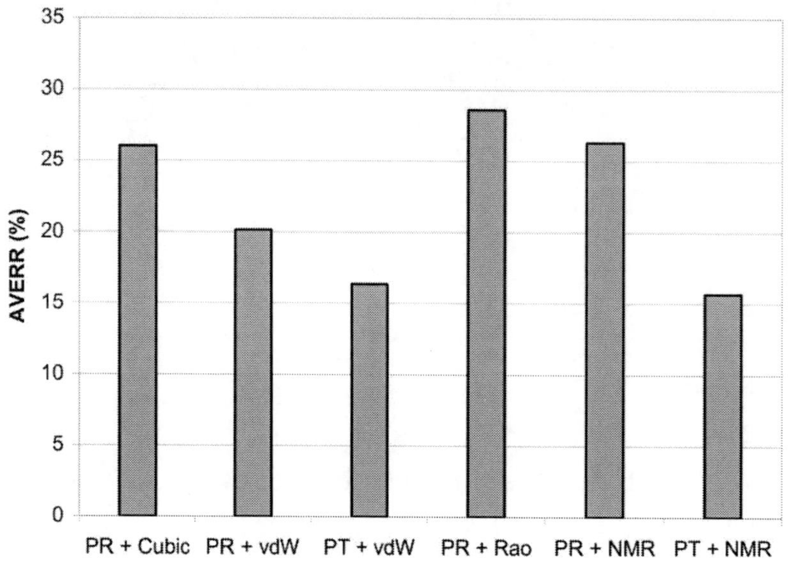

Figure 5. Comparison of the Different Thermodynamic Models

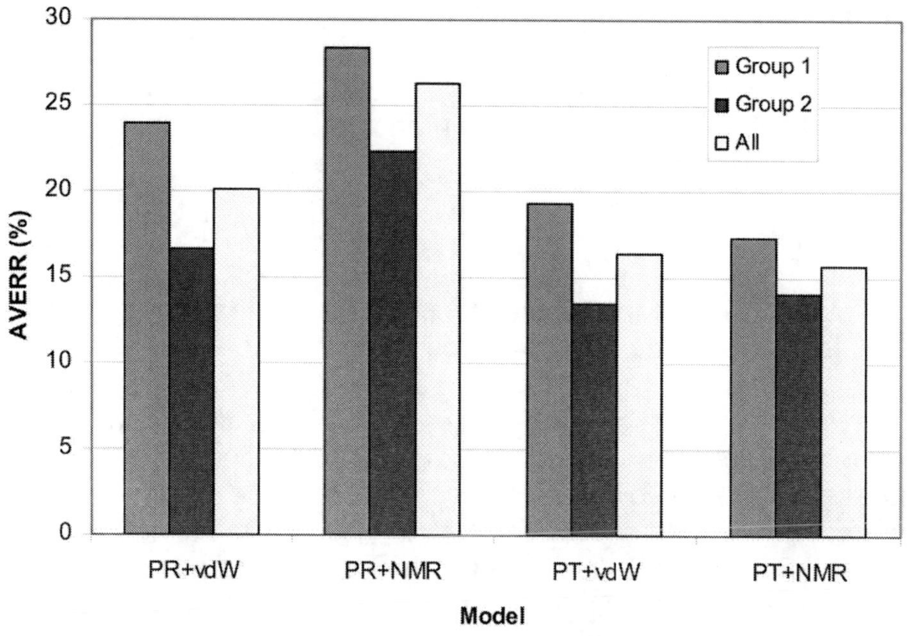

Figure 6. Summary of Errors by Model and by Data Group.

Predictive Capacity

It is interesting to compare the predictive capacity of the various models. To this end, Figure 7 has been prepared. It shows the experimental solubility of acridine in carbon dioxide at 318.15 [K] and the predictions using various models. The Patel-Teja EoS with the new mixing rule works best in this task. Interestingly, all models in that Figure have a similar capacity at high pressures. However, at low pressures, the Patel-Teja + NMR model is able to fit better the experimental data, although it underestimates them. The rest of the models overestimate the solubility at low pressures.

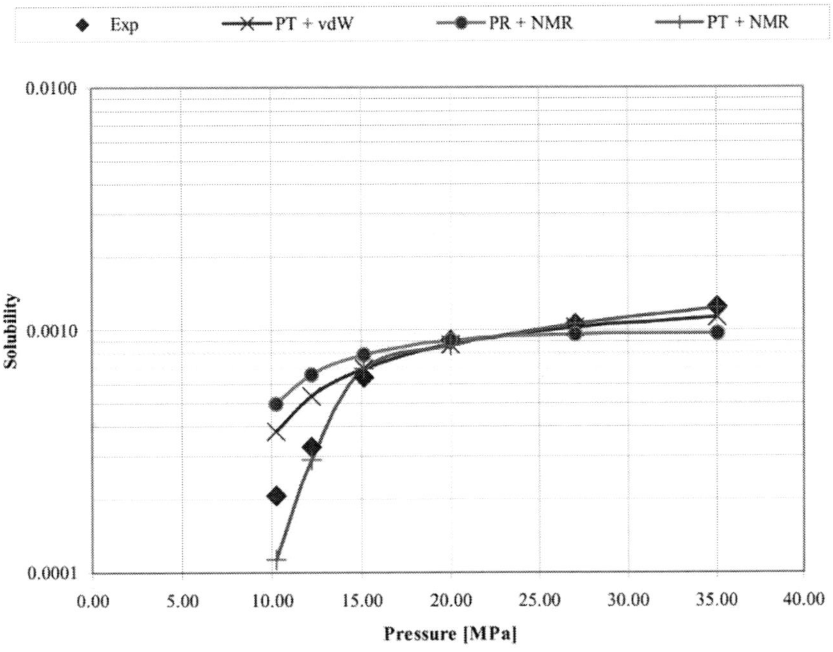

Figure 7. Solubility of Acridine in Carbon Dioxide

CONCLUSIONS

A new combination rule has been proposed to be used in the modeling and estimation of solubility of solids in supercritical fluids. The proposed rule is a modification of the classical van der Waals mixing rules where the binary cohesive parameter of the EoS (any EoS) is allowed to vary with pressure. Correlations for the proposed mixing rule were developed for the Peng-Robinson and Patel-Teja EoS based on the experimental data of 63 isotherms. The correlations were then applied to the rest of the isotherms with very good results.

Different thermodynamic models were compared, using different cubic equations of state, such as Peng-Robinson and Patel-Teja, and different mixing rules such as van der Waals, Rao, and cubic. These models are the most widely used in the calculation of the solubility of solids in supercritical fluids. The model that best fits the experimental values is a combination of the Patel-Teja EoS and the new (modified) van der Waals mixing rule.

REFERENCES[1]

Ashour, I.; Almehaideb, R.; Fateen S.; Aly, G., 2000. Representation of Solid-Supercritical Fluid Phase Equilibria Using Cubic Equations of State. *Fluid Phase Equilibria,* 167, 41-61.

Caballero, A. C.; Estévez, L. A., 1991. A Comparative Study of Mixing Rules in the Prediction of Solid-Vapor Equilibria. *Ind. Eng. Chem.,* Res. 30, 601-605.

Caballero, A. C.; Hernández L. N.; Estévez, L. A., 1992. Calculation of Interaction Parameters for Binary Solid-SCF Equilibria Using Several EOS and Mixing Rules. *J. Supercritical Fluids*, 5, 283-295.

Coutsikos, P.; Magoulas, K.; Kontogeorgis, G. M., 2003. Prediction of Solid-Gas Equilibria with the Peng Robinson Equation of State. *J. Supercritical Fluids*, 25, 197-212.

Huang, C.; Tang, M.; Tao, W.; Chen, Y., 2001. Calculation of the Solid Solubilities in Supercritical Carbon Dioxide Using a Modified Mixing Model. *Fluid Phase Equilibria*, 179, 67 - 84.

Krukonis, V. About Supercritical Fluids. Supercritical Fluids: Their Proliferation in the Pharma Industry. *http://www.phasex4scf.com/supercritical_fluids/about_supercritical_fluids.htm.* Accessed on March 12, 2007.

Patel, N. C.; Teja, A. S. 1982. A New Cubic Equation of State for Fluids and Fluid Mixtures. *Chem. Eng. Sci.*, 37, 463.

Peng, D. Y.; Robinson, D. B., 1976. A New Two-Constant Equation of State. *Ind. Eng. Chem. Fundam.*, 15, 59.

Pongsiri, N.; Viswanath, D. S. A. Comparative Study of Mixing Rules in the Prediction of Solid-Vapor Equilibria. 1989. *Ind. Eng. Chem. Res.*,28, 1918-1921.

Valderrama J. O.; Arce P. F.; Aznar, M., 2002. A New Non-Quadratic Mixing Rule and Its Application to Mixture Containing Supercritical CO_2. *International Society for the Advancement of Supercritical Fluids.*

[1] The short list of references included here contains the main ones used in the preparation of this article although they have not specifically been cited in the text.

Quantitative Measurements of Colloid-Surface Interactions from Microscopic Imaging and Inverse Density Functional Theory

Mingqing Lu[1], Richard Beckham[1], Michael A. Bevan[1], and David M. Ford[2]. (1) Department of Chemical Engineering, Texas A&M University, TAMU 3122, College Station, TX 77843-3122, (2) Department of Chemical Engineering, University of Massachusetts, Amherst, 209 Goessmann Laboratory, 686 North Pleasant Street, Amherst, MA 01003

We present a modeling technique, based on an inversion of classical density functional theory (DFT), that allows us to determine the potential energy of a single colloidal particle with a surface, based on particle density profiles measured in a dense colloidal system. The equilibrium particle density profiles are obtained with a unique combination of total internal reflection and confocal scanning laser microscopy, proposed by Wu and Bevan under the name of Diffusing Colloidal Probe Microscopy (DCPM). In this paper, the authors' previous inverse DFT work on hard spheres [Lu et al., J. Chem. Phys. 122, 224710 (2005)] is extended to soft particle-particle interaction potentials characteristic of many colloidal fluids. Both simulated (Monte Carlo) and real DCPM data were used to test the approach. For model screened electrostatic, van der waals, and depletion systems, we found that the inversion procedure reproduces the true particle-surface potential energy to an accuracy within typical DCPM experimental limitations (~0.1 kT) at low to moderate bulk colloidal densities. Sedimentation equilibrium of colloidal suspensions is also studied. The choice of DFT closures for different interaction potential was obtained. For the DCPM experiments, some numerical challenges were overcome in the conversion of raw image data to particle density profiles before applying the inverse DFT. This combination of DCPM and DFT holds great promise for rapid mapping of O(kT) surface energetics in a variety of nano- and bio-technology applications.

Phase Behavior of Polydisperse Branched Polyolefins Using the PC-SAFT Equation of State

Shekhar Jain, Aleksandra Dominik, and **Walter G. Chapman**. Department of Chemical and Biomolecular Engineering, Rice University, 6100 Main St., Houston, TX 77096

The physical properties of polymeric materials can be tailored to a customer's specific needs by the choice of the R-olefin comonomer and the adjustment of its content in the chain. Processing of such polymers requires an understanding of the phase behavior as a function of comonomer content as well as polydispersity. In this work we present a simple thermodynamic model for branched polyolefin systems in the framework of the PC-SAFT equation of state. The model efficiently captures the effect of chain branching, comonomer type, and branching density on the phase behavior of polyolefin solutions. Since few adjustable parameters are required, it can be applied to systems for which little experimental data is available. The model parameters follow sensible trends, providing robustness for extrapolations and parameter prediction. We use the model to demonstrate the effect of polydispersity in molecular weight and polydispersity in comonomer content on the phase behavior of polymer solutions. Finally, the simple formulation facilitates its incorporation into commercial software for phase equilibria calculations.

Interfacial Properties and Structure of Polymeric Blends and Solutions from Interfacial-SAFT (iSAFT) Density Functional Theory

Shekhar Jain, Aleksandra Dominik, Adam S. Bymaster, and Walter G. Chapman. Department of Chemical and Biomolecular Engineering, Rice University, 6100 Main St., Houston, TX 77005

Interfacial properties play an essential role in applications of great industrial importance, such as adhesion, polymer coating and thin films, paints, separations, catalysis, wettability or emulsions. This has motivated studies of complex macromolecular systems (hydrocarbons, proteins, polymers) in confined geometries, and at solid-fluid and fluid-fluid interfaces. As properties of complex chemical systems are being tailored at the nanoscale, a molecular level understanding of the interplay between interfacial properties, fluid structure, and macroscopic properties of the material is necessary. Experimental investigation of interfacial phenomena is hampered by the small scale of the systems. Molecular simulation provides an alternative to experimental techniques, but is computationally expensive, and thus often limited to short-chain molecules. Consequently, substantial efforts have been directed towards developing a theory for structure and thermodynamics of inhomogeneous polymeric solutions and blends

We have developed a new density functional theory (DFT) to describe phase behavior and microstructure of mixtures of polymeric fluids in inhomogeneous environments. The theory is based on Wertheim's thermodynamic perturbation theory for association; it treats the polyatomic system as a mixture of strongly associating monomers, and invokes the limit of complete association in the free energy functional. The theory is formulated in terms of segment density; therefore it offers the simplicity and computational efficiency of an atomic DFT. This feature of iSAFT makes it best suited for modeling interfacial properties and structure of chains of several hundreds or even thousands of segments. We demonstrate the success of the theory in capturing the effects of size and shape of the fluid molecules and system parameters such as fluid-fluid and surface-fluid interaction strengths on fluid structure and interfacial properties. Our DFT shares a common basis with the Statistical Associating Fluid Theory (SAFT), one of the most accurate and widely used theories for bulk polymer solutions. This self-consistency of the bulk and interfacial models is crucial; it enables us to predict the interfacial properties of the considered systems with a set of molecular parameters obtained from the readily available bulk properties. We compare results from iSAFT to molecular simulations, as well as experimental data for interfacial tension and molecular structure of realistic systems comprising polymer melts, blends and solutions.

AUTHOR INDEX

American Institute of Chemical Engineers
3 Park Avenue
New York, NY 10016-5991

ISBN: 978-1-60423-362-9